T0234754

SpringerBriefs in Probability and Mathematical Statistics

SpringerBriefs present concise summaries of cutting-edge research and practical applications across a wide spectrum of fields. Featuring compact volumes of 50 to 125 pages, the series covers a range of content from professional to academic. Briefs are characterized by fast, global electronic dissemination, standard publishing contracts, standardized manuscript preparation and formatting guidelines, and expedited production schedules.

Typical topics might include:

- A timely report of state-of-the art techniques
- A bridge between new research results, as published in journal articles, and a contextual literature review
- A snapshot of a hot or emerging topic
- Lecture of seminar notes making a specialist topic accessible for non-specialist readers
- SpringerBriefs in Probability and Mathematical Statistics showcase topics of current relevance in the field of probability and mathematical statistics

Manuscripts presenting new results in a classical field, new field, or an emerging topic, or bridges between new results and already published works, are encouraged. This series is intended for mathematicians and other scientists with interest in probability and mathematical statistics. All volumes published in this series undergo a thorough refereeing process.

The SBPMS series is published under the auspices of the Bernoulli Society for Mathematical Statistics and Probability.

More information about this series at http://www.springer.com/series/14353

J. Adolfo Minjárez-Sosa

Zero-Sum Discrete-Time Markov Games with Unknown Disturbance Distribution

Discounted and Average Criteria

J. Adolfo Minjárez-Sosa
Department of Mathematics
University of Sonora
Hermosillo, Sonora, Mexico

ISSN 2365-4333   ISSN 2365-4341   (electronic)
SpringerBriefs in Probability and Mathematical Statistics
ISBN 978-3-030-35719-1   ISBN 978-3-030-35720-7   (eBook)
https://doi.org/10.1007/978-3-030-35720-7

Mathematics Subject Classification: 91A15, 90C40, 62G05

This Springer imprint is published by the registered company Springer Nature Switzerland AG.
The registered company address is: Gewerbestrasse 11, 6330 Cham, Switzerland

To my two women: Francisca and Camila

Preface

Discrete-time zero-sum Markov games constitute a class of stochastic games introduced by Shapley in [65] whose evolution over time can be described as follows. At each stage, players 1 and 2 observe the current state x of the game and independently choose actions a and b, respectively. Then, player 1 receives a payoff $r(x,a,b)$ from player 2 and the game moves to a new state y in accordance with a transition probability or a transition function F as in (1), below. The payoffs are accumulated throughout the evolution of the game in a finite or infinite horizon under a specific optimality criterion.

Even though there are now many studies in this field under multiple variants, it is mostly assumed that all components of the game are completely known by the players. However, the environment itself in which it evolves could make this assumption unrealistic or too strong. Hence, the availability of approximation and estimation algorithms that provide players with some insights on the evolution of the game is important, so that they can select their actions more accurately.

An important feature of this book is that it will deal with a class of Markov games with Borel state and action spaces, and possibly unbounded payoffs, under discounted and average criteria, whose state process $\{x_t\}$ evolves according to a stochastic difference equation of the form

$$x_{t+1} = F(x_t, a_t, b_t, \xi_t), \quad t = 0, 1, \ldots \tag{1}$$

Here, the pair (a_t, b_t) represents the actions chosen by players 1 and 2, respectively, at time t, and $\{\xi_t\}$ is the disturbance process which is an observable sequence of independent and identically distributed random variables with *unknown* distribution θ for both players. In this scenario, our concern is in a game played over an infinite horizon evolving as follows. At stage t, once the players have observed the state x_t, and before choosing the actions a_t and b_t, players 1 and 2 implement a statistical estimation process to obtain estimates θ_t^1 and θ_t^2 of θ, respectively. Then, independently, the players adapt their decisions to such estimators to select actions $a = a_t(\theta_t^1)$ and $b = b_t(\theta_t^2)$. Next the game jumps to a new state according to the transition probability determined by Eq. (1) and the unknown distribution θ, and the process is repeated over and over again.

This book is the first part of a project whose objective is to make a systematic analysis on recent developments in this kind of games. Specifically, in this first part we will provide the theoretical foundations on the procedures combining statistical estimation and control techniques for the construction of strategies of the players. We generically call this combination "estimation and control" procedures. The second part of the project will deal with another class of games models, as well as with approximation and computational aspects.

The statistical estimation process will be studied from two approaches. In the first one, we assume that the distribution θ has a density ρ on $\Re^k$. In this case, there is a vast literature (see, e.g., [9–11, 27] and references therein) that provides different density estimation methods that might be easily adapted to the conditions imposed by the problem being analyzed. Among these we can mention kernel density estimation, L_q estimation for $q \geq 1$, and projection estimation, through which it is possible to obtain several important properties such as the rate of convergence. The second approach is provided by the empirical distribution θ_t defined by the random disturbance process $\{\xi_t\}$. This method is very general in the sense that both the random variables ξ_t and the distribution θ can be arbitrary. The price that must be paid due to this generality is that its applicability is restricted because it is necessary to impose stronger conditions than those of the previous case on the game model. Anyhow, the use of the empirical distribution has the additional advantage that it provides an approximation method of the value of the game and optimal strategies for players, in cases where the distribution θ is difficult to handle, by replacing θ with a simpler distribution given by θ_t. In general terms, our approach to obtain estimation and control procedures for both discounted and average criteria consists of combining a statistical estimation method suitable for θ with game theory techniques. Our starting point is to, first, prove the existence of a value of the game as well as measurable minimizers/maximizers in the Shapley equation. To this end, some conditions are imposed on the game model which fall within the weighted-norm approach proposed by Wessels in [76] and then fully studied in [23, 24, 31] for Markov decision processes (MDPs) and recently for zero-sum stochastic games in [32, 40, 41, 44, 48]. Thereby, the estimation method is adapted to these conditions to obtain appropriate convergence properties.

Clearly, the good behavior of the strategies obtained through the estimation and control procedures depends on the accuracy of the estimation method, and even more on the optimality criterion with which their performance is measured. For instance, it is well known that the discounted criterion strongly depends on the decisions selected in the early stages of the game, just where the estimation process yields deficient information about the unknown distribution θ. So, neither player

1 nor player 2 can generally ensure the existence of discounted optimal strategies. Hence the optimality under a discounted criterion is studied in an asymptotic sense. The notion of asymptotic optimality used in this book for Markov games was motivated from Schäl [67], who introduced this concept to study adaptive MDPs. In contrast, in view of the necessary asymptotic analysis in the study of the average criterion, the strategies obtained by means of estimation and control procedures turn out to be average optimal, providing suitable ergodicity conditions.

According to the historical development of the theories of stochastic control and Markov games, the problem of estimation and control for MDPs, also known as an adaptive Markov control problem, has received considerable attention in recent years (see, e.g., [2, 7, 22, 25, 26, 28, 29, 33–35, 52–55, 67] and references therein). In fact, even though approximation algorithms for stochastic games and games with partial information have been studied from several points of view (see, e.g., [8, 17, 20, 43, 46, 59, 60, 63], and references therein), in the field of statistical estimation and control procedures for Markov games the literature remains scarce; we can cite, for instance, [50, 56–58, 69, 70]. In particular, [56] deals with semi-Markov zero-sum games with unknown sojourn time distribution. The works [69, 70] study repeated games assuming that the transition law depends on an unknown parameter which is estimated by the maximum likelihood method, whereas [50, 56–58] deal with the theory developed in the context of this book.

The book is organized as follows. In Chap. 1 the class of Markov game models we deal with is introduced, together with the main elements necessary to define the game problem. Chapters 2 and 3 are devoted to analyze the discounted and the average criteria, respectively, where estimation and control procedures are presented under the assumption that the distribution θ has a density on $\mathfrak{R}^k$. Empirical estimation-approximation methods are given in Chap. 4. In this case, by using the empirical distribution to estimate θ both discounted and average criteria are analyzed. Finally, several examples of the class of Markov games studied throughout the book are given in Chap. 5. In this part we focus, mainly, on illustrating our assumptions on the game model, as well as on the numerical implementation of the estimation and control algorithm in specific examples.

Acknowledgments. The work of the author has been partially supported by Consejo Nacional de Ciencia y Tecnología (CONACYT-México) grant CB/2015-254306. Special thanks to Onésimo Hernández-Lerma who has read a draft version; his valuable comments and suggestions helped to improve the book. I also want to thank my colleagues Fernando Luque-Vásquez, Oscar Vega-Amaya, and Carmen Higuera-Chan, with whom I form the "controllers team" of the University of Sonora; undoubtedly, many of the ideas discussed in our long talks are included in this book. Finally, I deeply thank Donna Chernyk, Associate Editor at Springer, for her help.

Hermosillo, Mexico   J. Adolfo Minjárez-Sosa
August 2019

Contents

Summary of Notation and Terminology

Symbols and Abbreviations

$\mathbb{N}$	Set of positive integers
$\mathbb{N}_0$	Set of nonnegative integers
$\mathfrak{R}$	Set of real numbers
$\mathfrak{R}^+$	Set of nonnegative real numbers
$1_D(\cdot)$	Indicator function of the set D
:=	Equality by definition
a.e.	Almost everywhere
a.s.	Almost surely
i.i.d.	Independent and identically distributed
r.v.	Random variable
p.m.	Probability measure
l.s.c.	Lower semicontinuous
u.s.c.	Upper semicontinuous

Spaces of Functions

- The space $L_q = L_q(\mathfrak{R}^k)$, for $1 \leq q < \infty$, consists of all real-valued measurable functions on $\mathfrak{R}^k$ with finite L_q-norm:

$$\|\rho\|_{L_q} := \left(\int_{\mathfrak{R}^k} |\rho|^q \, d\mu \right)^{1/q}$$

with respect to the Lebesgue measure μ.
- A Borel space is a Borel subset of a complete separable metric space.

For a Borel space X, we use the following notation:

$\mathscr{B}(X)$ — Borel σ-algebra in X, and "measurable," for either sets or functions, means "Borel measurable."

$\mathbb{B}(X)$ — Space of real-valued bounded measurable functions on X with the supremum norm: $\|v\|_B := \sup_{x \in X} |v(x)|$.

$\mathbb{C}(X) \subset \mathbb{B}(X)$ — Subspace of bounded continuous functions.

$\mathbb{L}(X)$ — Space of lower semicontinuous functions and bounded from below.

$\mathbb{B}_W(X)$ — For a function $W : X \to [1, \infty)$, space of measurable functions with finite weighted norm (W-norm):
$$\|v\|_W := \sup_{x \in X} \frac{|v(x)|}{W(x)}.$$

$\mathbb{C}_W(X) \subset \mathbb{B}_W(X)$ — Subspace of W-bounded continuous functions.

$\mathbb{L}_W(X) \subset \mathbb{B}_W(X)$ — Subspace of W-bounded lower semicontinuous functions.

$\mathbb{P}(X)$ — Space of probability measures on X endowed with the weak topology (see Appendix B).

$\mathbb{P}(X|Y)$ — Family of stochastic kernels on X given Y, where X and Y are Borel spaces.

Chapter 1
Zero-Sum Markov Games

In this chapter we present the class of zero-sum Markov games we are interested in. We first introduce the game model which is a collection of objects describing the evolution in time of the games. In addition, in order to define the corresponding game problem, the concept of *strategies* for players is given along with the optimality criteria that will be analyzed in the next chapters.

We pay special attention to a class of Markov games whose evolution over time is modeled by means of a stochastic difference equation. It is precisely in this kind of games that we will study estimation and control schemes, assuming that the random disturbance process involved in their dynamics has an unknown distribution.

1.1 Game Models

A zero-sum Markov game model is defined by the collection

$$\mathscr{GM} := (X, A, B, \mathbb{K}_A, \mathbb{K}_B, Q, r) \tag{1.1}$$

formed by:

(a)   A Borel space X called the *state space*.
(b)   Borel spaces A and B representing the *action sets* for players 1 and 2, respectively.
(c)   The *constraint sets* $\mathbb{K}_A$ and $\mathbb{K}_B$ which are assumed to be Borel subsets of $X \times A$ and $X \times B$, respectively. Moreover, for each $x \in X$, the x-sections

$$A(x) := \{a \in A : (x, a) \in \mathbb{K}_A\}$$

and

$$B(x) := \{b \in B : (x, a) \in \mathbb{K}_B\}$$

© The Author(s), under exclusive license to Springer Nature Switzerland AG 2020
J. A. Minjárez-Sosa, *Zero-Sum Discrete-Time Markov Games*
with Unknown Disturbance Distribution, SpringerBriefs in Probability
and Mathematical Statistics, https://doi.org/10.1007/978-3-030-35720-7_1

represent the admissible actions or control sets for players 1 and 2, respectively, and the set

$$\mathbb{K} = \{(x,a,b) : x \in X,\ a \in A(x),\ b \in B(x)\}$$

of admissible state-actions triplets is a Borel subset of $X \times A \times B$.

(d) A stochastic kernel $Q(\cdot|\cdot)$ on X given $\mathbb{K}$, called the *transition law*. That is, if $x \in X$ is the state of the game at time t, and the players 1 and 2 select actions $a \in A(x)$ and $b \in B(x)$, respectively, then $Q(\cdot|x,a,b)$ is the distribution of the next state of the game:

$$Q(D|x,a,b) = \Pr[x_{t+1} \in D|x_t = x, a_t = a, b_t = b],\quad D \in \mathscr{B}(X). \qquad (1.2)$$

(e) A measurable function $r : \mathbb{K} \to \mathfrak{R}$ that represents the one-stage payoff function.

   The game is played as follows. At each time $t \in \mathbb{N}_0$, the players observe the state of the game $x_t = x \in X$. Next, players 1 and 2 select, independently, actions $a_t = a \in A(x)$ and $b_t = b \in B(x)$ respectively. Then, player 1 receives a payoff $r(x,a,b)$ from player 2, and the game jumps to a new state $x_{t+1} = y \in X$ according to the transition law $Q(\cdot|x,a,b)$. Once the game is in the new state, the process is repeated. Therefore the goal of player 1 (player 2) is to maximize (minimize) his/her rewards (cost).

1.1.1 Difference-Equation Games: Estimation and Control

There are many situations where the evolution of the game is modeled by a stochastic difference equation of the form

$$x_{t+1} = F(x_t, a_t, b_t, \xi_t),\quad t \in \mathbb{N}_0, \qquad (1.3)$$

where $F : \mathbb{K} \times S \to X$ is a given measurable function and $\{\xi_t\}$ is a sequence of observable i.i.d. random variables defined on a probability space $(\Omega, \mathscr{F}, P)$, taking values in a Borel space S, with common distribution $\theta \in \mathbb{P}(S)$, and independent of the initial state x_0. In this case the transition law Q in (1.2) is determined by the function F and the distribution θ as (see Appendix C.1)

$$Q(D|x,a,b) = \theta(\{s \in S : F(x,a,b,s) \in D\})$$

$$= \int_S 1_D[F(x,a,b,s)]\theta(ds),\quad D \in \mathscr{B}(X), \qquad (1.4)$$

for $(x,a,b) \in \mathbb{K}$. If θ has a density ρ on $S = \mathfrak{R}^k$ with respect to Lebesgue measure, then Q takes the form

$$Q(D|x,a,b) = \int_{\mathfrak{R}^k} 1_D[F(x,a,b,s)]\rho(s)ds, \ D \in \mathscr{B}(X), (x,a,b) \in \mathbb{K}.$$

Typical examples of dynamics as (1.3) appear, for instance, in autoregressive and linear games models. In the first one the game's state process evolves according to an equation of the form

$$x_{t+1} = G(a_t,b_t)x_t + \xi_t \ \text{for } t \in \mathbb{N}_0,$$

with initial state x_0, state space $X = [0,\infty)$, actions sets $A(x) \subset A = \mathfrak{R}$, and $B(x) \subset B = \mathfrak{R}$ for $x \in X$, and $G : A \times B \to (0,\lambda]$ is a given function with $\lambda < 1$. In the linear models the dynamic of the game is

$$x_{t+1} = x_t + a_t + b_t + \xi_t \ \text{for } t \in \mathbb{N}_0,$$

with x_0 given, where $X = A = B = \mathfrak{R}$. Both models will be analyzed in Chap. 5 together with additional examples to illustrate the theory developed throughout the book.

The estimation and control procedures, proposed in next chapters, are established for this kind of games by assuming that the random disturbance process $\{\xi_t\}$ is observable with a common and *unknown* distribution $\theta \in \mathbb{P}(S)$. In this context, unlike the standard evolution of a Markov game, at each stage $t \in \mathbb{N}_0$, on the knowledge of the state $x_t = x$ and possibly the history of the game, before choosing the actions a_t and b_t, players 1 and 2 get estimates θ_t^1 and θ_t^2 of the unknown distribution θ, respectively, and adapt independently their strategies to select actions $a = a_t(\theta_t^1) \in A(x)$ and $b = b_t(\theta_t^2) \in B(x)$. Next, the game evolves as the standard case, i.e., player 1 receives a payoff $r(x,a,b)$ from player 2, and the system visits a new state $x_{t+1} \in X$ according to the transition law in (1.4).

Assuming observability of the process $\{\xi_t\}$ allows to implement statistical estimation methods of θ. Among them, the empirical distribution constitutes the most general method in the sense that both the disturbance space S and the distribution $\theta \in \mathbb{P}(S)$ can be arbitrary (see Appendix B.1). In the particular case where θ has a density ρ on $S = \mathfrak{R}^k$, the spectrum of statistical methods to estimate ρ is extended, so that there are more options to choose the most appropriate according to the conditions of the problem being analyzed (see Appendix D). Both approaches will be analyzed in the following chapters; and it is worth emphasizing that they differ in the type of conditions needed for their implementation, and therefore in the arguments used in the corresponding proofs.

1.2 Strategies

The players select the actions by means of rules called strategies defined below.

Let $H_0 := X$ and $H_t := \mathbb{K} \times H_{t-1}$ for $t \in \mathbb{N}$. Then, a generic element of H_t is

$$h_t := (x_0, a_0, b_0, \ldots, x_{t-1}, a_{t-1}, b_{t-1}, x_t)$$

which represents the history of the game up to time t. On the other hand, for each $x \in X$, we define $\mathbb{A}(x) := \mathbb{P}(A(x))$ and $\mathbb{B}(x) := \mathbb{P}(B(x))$, as well as the sets of stochastic kernels

$$\Phi^1 := \left\{ \varphi^1 \in \mathbb{P}(A|X) : \varphi^1(\cdot|x) \in \mathbb{A}(x) \ \forall x \in X \right\}$$

$$\Phi^2 := \left\{ \varphi^2 \in \mathbb{P}(B|X) : \varphi^2(\cdot|x) \in \mathbb{B}(x) \ \forall x \in X \right\}.$$

Definition 1.1. (a) A *strategy* for player 1 is a sequence $\pi^1 = \{\pi^1_t\}$ of stochastic kernels $\pi^1_t \in \mathbb{P}(A|H_t)$ such that

$$\pi^1_t(A(x_t)|h_t) = 1 \ \forall h_t \in H_t, t \in \mathbb{N}_0.$$

We denote by Π^1 the family of all strategies for player 1.

(b) A strategy $\pi^1 = \{\pi^1_t\} \in \Pi^1$ is called a *Markov strategy* if π^1_t is in Φ^1 for all $t \in \mathbb{N}_0$, and it is called stationary if

$$\pi^1_t(\cdot|h_t) = \varphi^1(\cdot|x_t) \ \forall h_t \in H_t, t \in \mathbb{N}_0,$$

for some stochastic kernel φ^1 in Φ^1, so that π^1 is of the form $\pi^1 = \{\varphi^1, \varphi^1, \ldots\} := \{\varphi^1\}$.

We denote by Π^1_s the class of stationary strategies for player 1.

Let $\mathbb{F}^1$ be the set of all measurable functions $f^1 : X \to A$ such that $f^1(x) \in A(x)$ for all $x \in X$, and let $\mathbb{F}^2$ be the set of all measurable functions $f^2 : X \to B$ such that $f^2(x) \in B(x)$ for all $x \in X$.

Definition 1.2. A *strategy* $\pi^1 = \{\pi^1_t\}$ for player 1 is said to be a

(a) *pure (or deterministic)* strategy if there exists a sequence $\{g^1_t\}$ of measurable functions $g^1_t : H_t \to A$ such that, for all $h_t \in H_t$ and $t \in \mathbb{N}_0$, $g^1_t(h_t) \in A(x_t)$ and

$$\pi^1_t(A'|h_t) = 1_{A'}\left[g^1_t(h_t)\right] \quad \text{for all } A' \in \mathscr{B}(A),$$

which means that $\pi_t(\cdot|h_t)$ is concentrated at $g^1_t(h_t)$;

(b) *pure Markov* strategy if there is a sequence $\{f_t^1\}$ of functions $f_t^1 \in \mathbb{F}^1$ such that $\pi_t(\cdot|h_t)$ is concentrated at $f_t^1(x_t) \in A(x_t)$ for all $h_t \in H_t$ and $t \in \mathbb{N}_0$;

(c) *pure stationary* strategy if there is a function $f^1 \in \mathbb{F}^1$ such that $\pi_t(\cdot|h_t)$ is concentrated at $f^1(x_t) \in A(x_t)$ for all $h_t \in H_t$ and $t \in \mathbb{N}_0$.

We denote by Π_D^1 the family of all pure strategies for player 1.

The sets Π^2, Π_s^2, and Π_D^2 of all strategies, all stationary strategies, and pure strategies for player 2 are defined similarly. Observe that

$$\Pi_D^i \subset \Pi^i, \text{ for } i = 1, 2.$$

Wherever appropriate, we shall use the following notation related to the probability measures in the sets $\mathbb{A}(x)$ and $\mathbb{B}(x)$. For probability measures $\varphi^1(\cdot|x) \in \mathbb{A}(x)$ and $\varphi^2(\cdot|x) \in \mathbb{B}(x)$, and $x \in X$, we write $\varphi^i(x) = \varphi^i(\cdot|x)$, $i = 1, 2$. In addition, for a measurable function $u : \mathbb{K} \to \mathfrak{R}$,

$$u(x, \varphi^1, \varphi^2) := \int_{B(x)} \int_{A(x)} u(x, a, b) \varphi^1(da|x) \varphi^2(db|x) = u(x, \varphi^1(x), \varphi^2(x)). \quad (1.5)$$

For instance, for $x \in X$ we have

$$r(x, \varphi^1, \varphi^2) := \int_{B(x)} \int_{A(x)} r(x, a, b) \varphi^1(da|x) \varphi^2(db|x).$$

For the case of games evolving according to a difference equation as (1.3), we consider histories of the form

$$h_t := (x_0, a_0, b_0, \xi_0, \dots, x_{t-1}, a_{t-1}, b_{t-1}, \xi_{t-1}, x_t) \in \mathbb{K} \times S \times H_{t-1}, \quad (1.6)$$

for $t \in \mathbb{N}$. In addition, for $x \in X$ and $s \in S$, we write

$$v(F(x, \varphi^1, \varphi^2, s)) := \int_{B(x)} \int_{A(x)} v((F(x, a, b, s)) \varphi^1(da|x) \varphi^2(db|x),$$

for a measurable function $v : X \to \mathfrak{R}$.

1.3 Markov Game State Process

Let $(\Omega', \mathscr{F}')$ be the measurable space consisting of the sample space $\Omega' = \mathbb{K}^\infty$ and its product σ-algebra $\mathscr{F}'$. Following standard arguments (see, e.g., [13]), from the Theorem of C. Ionescu-Tulcea (Proposition C.2 in Appendix C) we have that for each pair of strategies $(\pi^1, \pi^2) \in \Pi^1 \times \Pi^2$ and initial state $x_0 = x \in X$, there exists a unique probability measure $P_x^{\pi^1, \pi^2}$ and a stochastic process $\{(x_t, a_t, b_t)\}$, where x_t

and (a_t, b_t) represent the state and the actions of the players, respectively, at stage $t \in \mathbb{N}_0$, satisfying, for $D \in \mathscr{B}(X)$, $A' \in \mathscr{B}(A)$, and $B' \in \mathscr{B}(B)$,

$$P_x^{\pi^1,\pi^2} \left[x_0 \in D \right] = \delta_x(D); \tag{1.7}$$

$$P_x^{\pi^1,\pi^2} \left[a_t \in A' | h_t \right] = \pi_t^1 \left(A' | h_t \right); \tag{1.8}$$

$$P_x^{\pi^1,\pi^2} \left[b_t \in B' | h_t \right] = \pi_t^2 \left(B' | h_t \right); \tag{1.9}$$

$$P_x^{\pi^1,\pi^2} \left[a_t \in A', b_t \in B' | h_t \right] = \pi_t^1 \left(A' | h_t \right) \pi_t^2 \left(B' | h_t \right); \tag{1.10}$$

$$P_x^{\pi^1,\pi^2} \left[x_{t+1} \in D | h_t, a_t, b_t \right] = Q \left(D | x_t, a_t, b_t \right); \tag{1.11}$$

where $\delta_x(\cdot)$ is the Dirac measure concentrated at x. We denote by $E_x^{\pi^1,\pi^2}$ the expectation operator with respect to $P_x^{\pi^1,\pi^2}$.

In the scenario of difference equation games (1.3), the measurable space consists of the sample space $\Omega' = (\mathbb{K} \times S)^\infty$ with the corresponding product σ-algebra $\mathscr{F}'$. Then, considering histories of the form (1.6), in addition to the properties (1.7)–(1.11), we have that for each $(\pi^1, \pi^2) \in \Pi^1 \times \Pi^2$ and $x \in X$, the probability measure $P_x^{\pi^1,\pi^2}$ and the stochastic process $\{(x_t, a_t, b_t, \xi_t)\}$, satisfy

$$P_x^{\pi^1,\pi^2} \left[\xi_t \in S' | h_t, a_t, b_t \right] = \theta(S'), \quad S' \in \mathscr{B}(S).$$

The stochastic process $\{x_t\}$ defined on $(\Omega', \mathscr{F}', P_x^{\pi^1,\pi^2})$ is called the *game's state process*.

1.4 Optimality Criteria

For each pair of strategies $(\pi^1, \pi^2) \in \Pi^1 \times \Pi^2$ and initial state $x_0 = x \in X$, we define the *total expected α-discounted payoff* as

$$V_\alpha(x, \pi^1, \pi^2) := E_x^{\pi^1,\pi^2} \left[\sum_{t=0}^\infty \alpha^t r(x_t, a_t, b_t) \right], \tag{1.12}$$

where $\alpha \in (0, 1)$ represents the discount factor. We also define the *long-run expected average payoff* as

$$J(x, \pi^1, \pi^2) := \liminf_{n \to \infty} \frac{1}{n} E_x^{\pi^1,\pi^2} \sum_{t=0}^{n-1} r(x_t, a_t, b_t). \tag{1.13}$$

The lower and the upper values of the discounted game are given as:

$$L_\alpha(x) := \sup_{\pi^1 \in \Pi^1} \inf_{\pi^2 \in \Pi^2} V_\alpha(x, \pi^1, \pi^2), \quad x \in X, \tag{1.14}$$

and

$$U_\alpha(x) := \inf_{\pi^2 \in \Pi^2} \sup_{\pi^1 \in \Pi^1} V_\alpha(x, \pi^1, \pi^2), \quad x \in X, \tag{1.15}$$

respectively. Observe that, in general, $U_\alpha(\cdot) \geq L_\alpha(\cdot)$, but if it holds that $U_\alpha(\cdot) = L_\alpha(\cdot)$, the common function is called *the α-value of the game* and is denoted by $V_\alpha(\cdot)$. Now, if the discounted game has a value $V_\alpha(\cdot)$, a strategy $\pi^1_* \in \Pi^1$ is said to be *α-optimal for player 1* if

$$V_\alpha(x) = \inf_{\pi^2 \in \Pi^2} V_\alpha(x, \pi^1_*, \pi^2), \quad x \in X.$$

Similarly, a strategy $\pi^2_* \in \Pi^2$ is said to be *α-optimal for the player 2* if

$$V_\alpha(x) = \sup_{\pi^1 \in \Pi^1} V_\alpha(x, \pi^1, \pi^2_*), \quad x \in X.$$

In this case, (π^1_*, π^2_*) is an *α-optimal pair* of strategies or *saddle point*. Note that (π^1_*, π^2_*) is an optimal pair if and only if

$$V_\alpha(x, \pi^1, \pi^2_*) \leq V_\alpha(x, \pi^1_*, \pi^2_*) \leq V_\alpha(x, \pi^1_*, \pi^2) \tag{1.16}$$

for all $x \in X, \pi_1 \in \Pi^1, \pi^2 \in \Pi^2$.

The lower value $L(\cdot)$ and the upper value $U(\cdot)$ for the average payoff criterion are defined similarly, and the *average value of the game* is denoted by $J(\cdot)$. Then, if the average game has a value $J(\cdot)$, a strategy $\pi^1_* \in \Pi^1$ is said to be *average optimal for player 1* if

$$J(x) = \inf_{\pi^2 \in \Pi^2} J(x, \pi^1_*, \pi^2), \quad x \in X; \tag{1.17}$$

and a strategy $\pi^2_* \in \Pi^2$ is said to be *average optimal for player 2* if

$$J(x) = \sup_{\pi^1 \in \Pi^1} J(x, \pi^1, \pi^2_*), \quad x \in X. \tag{1.18}$$

The pair (π^1_*, π^2_*) is called an *average optimal pair* of strategies if (1.17) and (1.18) hold; equivalently,

$$J(x, \pi^1, \pi^2_*) \leq J(x, \pi^1_*, \pi^2_*) \leq J(x, \pi^1_*, \pi^2) \tag{1.19}$$

for all $x \in X, \pi_1 \in \Pi^1, \pi^2 \in \Pi^2$.

ε-Optimal Strategies. If the α-discounted game has a value $V_\alpha(\cdot)$, a strategy $\pi^1_* \in \Pi^1$ is said to be *ε-optimal for player 1*, for $\varepsilon \geq 0$, if

$$V_\alpha(\cdot) - \varepsilon \leq \inf_{\pi^2 \in \Pi^2} V_\alpha(\cdot, \pi^1_*, \pi^2).$$

In addition, a strategy $\pi_*^2 \in \Pi^2$ is said to be ε-*optimal for the player 2* if

$$V_\alpha(\cdot) + \varepsilon \geq \sup_{\pi^1 \in \Pi^1} V_\alpha(\cdot, \pi^1, \pi_*^2).$$

Therefore, a strategy $\pi_*^i \in \Pi^i$ is optimal for player i, $i = 1, 2$, if it is 0-optimal.

In a similar way we define ε-optimal strategies for the average payoff criterion.

Our objective in the following chapters is to study the existence of optimal pairs of strategies for difference equation games (1.3) when the disturbance distribution θ is unknown for both players. Specifically, we will establish estimation and control procedures that lead to the construction of optimal strategies.

Chapter 2
Discounted Optimality Criterion

We consider the game model

$$\mathscr{GM} := (X, A, B, \mathbb{K}_A, \mathbb{K}_B, Q, r)$$

introduced in (1.1). The problems we are concerned with in this chapter are those related to the discounted case, which are summarized as follows.

1. Establish conditions to prove the existence of a value of the game and a pair of optimal strategies. That is, prove the existence of a function V_α and a pair of strategies $(\pi_*^1, \pi_*^2) \in \Pi^1 \times \Pi^2$ such that, for all $x \in X, (\pi^1, \pi^2) \in \Pi^1 \times \Pi^2$, $U_\alpha(x) = L_\alpha(x) = V_\alpha(x)$, and

$$V_\alpha(x, \pi^1, \pi_*^2) \leq V_\alpha(x, \pi_*^1, \pi_*^2) = V_\alpha(x) \leq V_\alpha(x, \pi_*^1, \pi^2),$$

where

$$V_\alpha(x, \pi^1, \pi^2) := E_x^{\pi^1, \pi^2}\left[\sum_{t=0}^{\infty} \alpha^t r(x_t, a_t, b_t)\right] \tag{2.1}$$

is the total expected α-discounted payoff, and U_α and L_α are the upper and the lower values of the game (see (1.12), (1.14), (1.15), and (1.16)). To obtain these results we need some concepts and techniques on multifunctions and measurable selectors that are summarized in Appendix A.

2. Once the previous problem is solved, we will focus on introducing estimation and control procedures in difference equation game models, described in Sect. 1.1.1, assuming that $\{\xi_t\}$ is a sequence of i.i.d. random variables with unknown density ρ.

© The Author(s), under exclusive license to Springer Nature Switzerland AG 2020
J. A. Minjárez-Sosa, *Zero-Sum Discrete-Time Markov Games with Unknown Disturbance Distribution*, SpringerBriefs in Probability and Mathematical Statistics, https://doi.org/10.1007/978-3-030-35720-7_2

2.1 Minimax-Maximin Optimality Conditions

Throughout the chapter we will be dealing with the following Shapley's operator defined on a family of measurable functions v on X:

$$T_\alpha v(x) := \inf_{\varphi^2 \in \mathbb{B}(x)} \sup_{\varphi^1 \in \mathbb{A}(x)} \left[r(x, \varphi^1, \varphi^2) + \alpha \int_X v(y) Q\left(dy|x, \varphi^1, \varphi^2\right) \right], \qquad (2.2)$$

for $x \in X$. The main issue in (2.2) is to ensure the interchange of inf and sup as well as the existence of measurable selectors of the multifunctions $\mathbb{A}(x)$ and $\mathbb{B}(x)$ (see Definition A.4 in Appendix A). To this end, it is needed to impose suitable continuity and compactness conditions on the game model such as the following.

Assumption 2.1 *(a) The multifunctions $x \to A(x)$ and $x \to B(x)$ are compact-valued and continuous.*

(b) The payoff function $r(x, a, b)$ is continuous in $(x, a, b) \in \mathbb{K}$.

(c) The transition law Q is weakly continuous, that is, the mapping

$$(x, a, b) \to \int_X v(y) Q\left(dy|x, a, b\right)$$

is continuous on $\mathbb{K}$, for every function $v \in \mathbb{C}(X)$.

Assumption 2.2 *(a) For each $x \in X$, the sets $A(x)$ and $B(x)$ are compact.*

(b) For each $(x, a, b) \in \mathbb{K}$, $r(x, \cdot, b)$ is upper semicontinuous (u.s.c.) on $A(x)$, and $r(x, a, \cdot)$ is lower semicontinuous (l.s.c.) on $B(x)$.

(c) For each $(x, a, b) \in \mathbb{K}$ and $v \in \mathbb{B}(X)$, the functions

$$a \to \int_X v(y) Q(dy|x, a, b) \quad and \quad b \to \int_X v(y) Q(dy|x, a, b)$$

are continuous on $A(x)$ and $B(x)$, respectively.

For a measurable function $v : X \to \mathfrak{R}$, $(x, a, b) \in \mathbb{K}$, $\varphi^1(x) \in \mathbb{A}(x)$, and $\varphi^2(x) \in \mathbb{B}(x)$, we define (see (1.5))

$$H(x, a, b) := r(x, a, b) + \alpha \int_X v(y) Q\left(dy|x, a, b\right)$$

and

$$\bar{H}(x, \varphi^1, \varphi^2) := \int_{B(x)} \int_{A(x)} H(x, a, b) \varphi^1(da|x) \varphi^2(db|x).$$

From "extended Fatou Lemma" (see [31, Lemma 8.3.7]) and Proposition B.2 in Appendix B, it is easy to prove that if H is l.s.c. (u.s.c.) on $\mathbb{K}$, then $\bar{H}$ is also l.s.c. (u.s.c.) on $X \times \mathbb{A}(x) \times \mathbb{B}(x)$. Thus, from Proposition B.3, provided that either Assumption 2.1 or 2.2 holds, Berge's Theorem, Fan's minimax theorem, and suitable selection theorems, given in Appendix A (see also [61]), yield the existence of $(\varphi_*^1, \varphi_*^2) \in \Phi^1 \times \Phi^2$, such that, for all $x \in X$, $\varphi_*^1(x) \in \mathbb{A}(x)$ and $\varphi_*^2(x) \in \mathbb{B}(x)$ satisfy

$$
\begin{aligned}
T_\alpha v(x) &= \sup_{\varphi^1 \in \mathbb{A}(x)} \inf_{\varphi^2 \in \mathbb{B}(x)} \left[r(x, \varphi^1, \varphi^2) + \alpha \int_X v(y) Q\left(dy|x, \varphi^1, \varphi^2\right) \right] \\
&= r(x, \varphi_*^1, \varphi_*^2) + \alpha \int_X v(y) Q(dy|x, \varphi_*^1, \varphi_*^2) \\
&= \max_{\varphi^1 \in \mathbb{A}(x)} \left[r(x, \varphi^1, \varphi_*^2) + \alpha \int_X v(y) Q(dy|x, \varphi^1, \varphi_*^2) \right] \\
&= \min_{\varphi^2 \in \mathbb{B}(x)} \left[r(x, \varphi_*^1, \varphi^2) + \alpha \int_X v(y) Q(dy|x, \varphi_*^1, \varphi^2)) \right].
\end{aligned}
\tag{2.3}
$$

It is worth observing that if the continuity, convexity, and concavity conditions, requested in Berge's Theorem and Fan's minimax Theorem, are satisfied by the components of the game model, then either Assumption 2.1 or 2.2, together a selection theorem (see Appendix A), yield the existence of $(f_*^1, f_*^2) \in \mathbb{F}^1 \times \mathbb{F}^2$, such that for all $x \in X$, $f_*^1(x) \in A(x)$ and $f_*^2(x) \in B(x)$ satisfy

$$
\begin{aligned}
T_\alpha v(x) &= \sup_{a \in A(x)} \inf_{b \in B(x)} \left[r(x, a, b) + \alpha \int_X v(y) Q\left(dy|x, a, b\right) \right] \\
&= r(x, f_*^1, f_*^2) + \alpha \int_X v(y) Q(dy|x, f_*^1, f_*^2) \\
&= \max_{a \in A(x)} \left[r(x, a, f_*^2) + \alpha \int_X v(y) Q(dy|x, a, f_*^2) \right] \\
&= \min_{b \in B(x)} \left[r(x, f_*^1, b) + \alpha \int_X v(y) Q(dy|x, f_*^1, b)) \right].
\end{aligned}
\tag{2.4}
$$

This situation occurs, for instance, in the linear-quadratic games as we will see in Chap. 5.

Another set of minimax (maximin) optimality conditions implying similar relations to (2.3) is the following weaker assumption.

Assumption 2.3 *(a) The mapping $x \to A(x)$ is l.s.c. and $A(x)$ is complete for every $x \in X$;*

(b) The mapping $x \to B(x)$ is u.s.c. and $B(x)$ is compact for every $x \in X$;

(c) The function $r(\cdot, \cdot, \cdot) \geq 0$ is l.s.c. on $\mathbb{K}$;

(d) The mapping

$$(x,a,b) \to \int_X v(y)Q(dy|x,a,b) \tag{2.5}$$

is continuous on $\mathbb{K}$ for every $v \in \mathbb{C}(X)$.

Note that from Proposition C.1(b), Assumption 2.3(d) (or Assumption 2.1(c)) implies that the mapping in (2.5) is lower semicontinuous whenever $v(\cdot)$ is in $\mathbb{L}(X)$.

Remark 2.1. Under Assumption 2.3, from Küenle [47] we have that the operator T_α maps $\mathbb{L}(X)$ into itself. In addition, for all $x \in X$ and $v \in \mathbb{L}(X)$

$$T_\alpha v(x) := \sup_{\varphi^1 \in \mathbb{A}(x)} \inf_{\varphi^2 \in \mathbb{B}(x)} \left[r(x, \varphi^1, \varphi^2) + \alpha \int_X v(y)Q(dy|x, \varphi^1, \varphi^2) \right]. \tag{2.6}$$

Moreover, for each $\varepsilon > 0$, there exist $\varphi_\varepsilon^1 \in \Phi_1$ and $\varphi_*^2 \in \Phi_2$ such that, for all $x \in X$,

$$T_\alpha v(x) = \sup_{\varphi^1 \in \mathbb{A}(x)} \left[r(x, \varphi^1, \varphi_*^2) + \alpha \int_X v(y)Q(dy|x, \varphi^1, \varphi_*^2) \right], \tag{2.7}$$

and

$$T_\alpha v(x) - \varepsilon \leq \inf_{\varphi^2 \in \mathbb{B}(x)} \left[r(x, \varphi_\varepsilon^1, \varphi^2) + \alpha \int_X v(y)Q(dy|x, \varphi_\varepsilon^1, \varphi^2) \right] \quad \forall x \in X. \tag{2.8}$$

2.2 Discounted Optimal Strategies

In this section we address the problem on the existence of the value function and optimal strategies for the discounted game. The theory is developed under the setting of the weaker Assumption 2.3 together with the following growth condition that allows us to handle the unbounded payoff case.

Assumption 2.4 *There exist a continuous function $W \geq 1$ on X and positive constants $\beta < 1$, $M < \infty$, and $d < \infty$ such that the following inequalities hold for all $(x,a,b) \in \mathbb{K}$:*

(a) $0 \leq r(x,a,b) \leq MW(x)$;

(b) $\int_X W(y)Q(dy|x,a,b) \leq \beta W(x) + d$.

(c) The function

$$(x,a,b) \to \int_X W(y)Q(dy|x,a,b)$$

is continuous on $\mathbb{K}$.

Remark 2.2. If the payoff function r is bounded, say by the constant M, then Assumption 2.4 holds by taking $W \equiv 1$ and $d = 1$.

From Van Nunen and Wessels [72], we have that Assumption 2.4 together with Assumption 2.3 implies that the Shapley's operator T_α has a unique fixed point in a certain measurable functions space. To state this fact precisely we first introduce some definitions and notation.

For each measurable function $u : X \to \mathfrak{R}$, define the *W-weighted norm*, W-norm for short, as

$$\|u\|_W := \sup_{x \in X} \frac{|u(x)|}{W(x)},$$

and denote by $\mathbb{B}_W$ the normed linear space of all measurable functions with finite W-norm. Let $\mathbb{L}_W$ be the subspace of functions in $\mathbb{L}(X)$ that belongs to $\mathbb{B}_W$. It is easy to verify that $\mathbb{L}_W$ is a Banach space.

Remark 2.3 (Contraction Property of T_α [72]). Suppose that Assumptions 2.3 and 2.4 hold. For each discount factor $\alpha \in (0,1)$, we fix an arbitrary number $\gamma_\alpha \in (\alpha, 1)$ and define the function $\bar{W}(x) := W(x) + e$, $x \in X$, where $e := d(\gamma_\alpha/\alpha - 1)^{-1}$. Consider the space $\mathbb{B}_{\bar{W}}$ of measurable functions $v : X \to \mathfrak{R}$ with finite $\bar{W}$-norm, that is

$$\|v\|_{\bar{W}} := \sup_{x \in X} \frac{|v(x)|}{\bar{W}(x)} < \infty.$$

Observe that $\mathbb{B}_W = \mathbb{B}_{\bar{W}}$ and the norms $\|\cdot\|_W$ and $\|\cdot\|_{\bar{W}}$ are equivalent since

$$\|v\|_{\bar{W}} \leq \|v\|_W \leq l_\alpha \|v\|_{\bar{W}} \quad \text{for } v \in \mathbb{B}_W, \tag{2.9}$$

where

$$l_\alpha := 1 + e = 1 + \frac{\alpha d}{\gamma_\alpha - \alpha}. \tag{2.10}$$

Then, from [72, Lemma 1], the function $\bar{W}$ satisfies the inequality

$$\alpha \int_S \bar{W}[F(x,a,b,s)]\mu(ds) \leq \gamma_\alpha \bar{W}(x) \ \forall (x,a,b) \in \mathbb{K}. \tag{2.11}$$

Following straightforward calculations, it is easy to see that, for each $\alpha \in (0,1)$, inequality (2.11) implies that operator T_α is a contraction from $\mathbb{L}_W$ into itself with respect to the $\bar{W}$-norm with modulus γ_α. Hence:
(a) For all $v, u \in \mathbb{B}_W$,

$$\|T_\alpha v - T_\alpha u\|_{\bar{W}} \leq \gamma_\alpha \|v - u\|_{\bar{W}}. \tag{2.12}$$

(b) Thus, by the Banach Fixed Point Theorem, T_α has a unique fixed point $V_\alpha \in \mathbb{L}_W$, i.e.,

$$T_\alpha V_\alpha = V_\alpha \tag{2.13}$$

and, as $n \to \infty$,

$$\|T_\alpha^n u - V_\alpha\|_{\bar{W}} \to 0 \quad \forall u \in \mathbb{L}_W,$$

where $T_\alpha^n u = T_\alpha(T_\alpha^{n-1} u)$ for $n \geq 1$.

(c) Furthermore, since $\|\cdot\|_W$ and $\|\cdot\|_{\bar{W}}$ are equivalent norms on $\mathbb{L}_W$,

$$\|T_\alpha^n u - V_\alpha\|_W \to 0 \text{ as } n \to \infty, \quad \forall u \in \mathbb{L}_W.$$

The next theorem shows the existence of optimal strategies for the discounted stochastic game. Its proof follows from (2.6) to (2.8), Remark 2.3, and standard dynamic programming arguments.

Theorem 2.5. *Suppose that Assumptions 2.3 and 2.4 hold. Then:*

(a) The function V_α in Remark 2.3 is the value of the discounted game;

(b) There exists $\varphi_*^2 \in \Phi^2$ such that

$$V_\alpha(x) = \sup_{\varphi^1 \in \mathbb{A}(x)} \left[r(x, \varphi^1, \varphi_*^2) + \alpha \int_X V_\alpha(y) Q(dy|x, \varphi^1, \varphi_*^2) \right] \quad \forall x \in X;$$

moreover, for each $\varepsilon > 0$ there exists $\varphi_\varepsilon^1 \in \Phi^1$ such that

$$V_\alpha(x) - \varepsilon \leq \inf_{\varphi^2 \in \mathbb{B}(x)} \left[r(x, \varphi_\varepsilon^1, \varphi^2) + \alpha \int_X V_\alpha(y) Q(dy|x, \varphi_\varepsilon^1, \varphi^2) \right] \quad \forall x \in X.$$

(c) Thus, $\pi_\varepsilon^1 = \{\varphi_\varepsilon^1\} \in \Pi_s^1$ is an ε-optimal strategy for player 1 and $\pi_*^2 = \{\varphi_*^2\} \in \Pi_s^2$ is an optimal strategy for player 2, that is,

$$V_\alpha(\cdot) - \varepsilon \leq \inf_{\pi^2 \in \Pi_2} V_\alpha(\cdot, \pi_\varepsilon^1, \pi^2)$$

$$V_\alpha(\cdot) = \sup_{\pi^1 \in \Pi^1} V_\alpha(\cdot, \pi^1, \pi_*^2).$$

Theorem 2.5 states the existence of a l.s.c. value function V_α. It is clear that if continuity of V_α is required, we need to impose more restrictive conditions. For instance, as is proved in [42, Theorem 4.2], this fact is achieved under Assumptions 2.1 and 2.4. Specifically, we have the following result.

Theorem 2.6. *Suppose that Assumptions 2.1 and 2.4 hold. Then, for each $\alpha \in (0, 1)$:*

(a) The discounted payoff game has a value $V_\alpha \in \mathbb{C}_W$ and

$$\|V_\alpha\|_W \leq \frac{M}{1 - \alpha}. \tag{2.14}$$

(b) The value V_α satisfies $T_\alpha V_\alpha = V_\alpha$, and there exists $(\varphi_*^1, \varphi_*^2) \in \Phi^1 \times \Phi^2$, such that $\varphi_*^1(x) \in \mathbb{A}(x)$ and $\varphi_*^2 \in \mathbb{B}(x)$ satisfy, for all $x \in X$,

$$V_\alpha(x) = r(x, \varphi_*^1, \varphi_*^2) + \alpha \int_X V_\alpha(y)Q(dy|x, \varphi_*^1, \varphi_*^2)$$

$$= \max_{\varphi^1 \in \mathbb{A}(x)} \left[r(x, \varphi^1, \varphi_*^2) + \alpha \int_X V_\alpha(y)Q(dy|x, \varphi^1, \varphi_*^2) \right] \qquad (2.15)$$

$$= \min_{\varphi^2 \in \mathbb{B}(x)} \left[r(x, \varphi_*^1, \varphi^2) + \alpha \int_X V_\alpha(y)Q(dy|x, \varphi_*^1, \varphi^2) \right]. \qquad (2.16)$$

In addition, $\pi_*^1 = \{\varphi_*^1\} \in \Pi_s^1$ and $\pi_*^2 = \{\varphi_*^2\} \in \Pi_s^2$ form an optimal pair of strategies.

From Remark 2.3, the condition in Assumption 2.4(b) can be replaced by the following assumption (see (2.11)).

Assumption 2.7 *There exists a constant $\tilde{\gamma} \in \left(1, \dfrac{1}{\alpha}\right)$ such that*

$$\int_S W(y)Q(dy|x, a, b, s) \le \tilde{\gamma}W(x), \quad \forall(x, a, b) \in \mathbb{K}.$$

Moreover, under Assumptions 2.4(a), (c) and 2.7, it is easy to prove that the operator T_α is a contraction with respect to the W-norm modulus $\alpha\tilde{\gamma} \in (0, 1)$. Hence, provided that Assumption 2.3 (or Assumption 2.1) also holds, Theorem 2.5 (or Theorem 2.6) remains valid.

2.2.1 Asymptotic Optimality

There are some situations in which it is not possible to obtain optimal strategies for players under the usual criterion. If this is the situation, it is necessary to search schemes that give strategies that are "nearly" optimal, which involves to analyze the concept of optimality in a weaker sense. This is the case that will be treated in the next section for difference equation games models, where a statistical estimation method for the unknown density of the random disturbance process is combined with control procedures in order to construct strategies. Specifically, we adapt to stochastic games the notion of *asymptotic optimality* introduced by Schäl in [67] (see also [28]) to study adaptive Markov control processes under the discounted criterion. To introduce this weaker optimality criterion, we define the *discrepancy function* $D : \mathbb{K} \to \mathfrak{R}$ as

$$D(x, a, b) := r(x, a, b) + \alpha \int_X V_\alpha(y)Q(dy|x, a, b) - V_\alpha(x)$$

for all $(x,a,b) \in \mathbb{K}$. Observe that the relation (2.13) is equivalent to

$$\sup_{\varphi^1 \in \mathbb{A}(x)} \inf_{\varphi^2 \in \mathbb{B}(x)} D(x, \varphi^1, \varphi^2) = \inf_{\varphi^2 \in \mathbb{B}(x)} \sup_{\varphi^1 \in \mathbb{A}(x)} D(x, \varphi^1, \varphi^2) = 0.$$

Moreover, from Theorem 2.5(b), the pair $(\varphi_\varepsilon^1, \varphi_*^2)$ satisfies, for all $x \in X$,

$$D(x, \varphi_\varepsilon^1, \varphi^2) \geq -\varepsilon \quad \forall \varphi^2 \in \mathbb{B}(x),$$
$$D(x, \varphi^1, \varphi_*^2) \leq 0 \quad \forall \varphi^1 \in \mathbb{A}(x). \tag{2.17}$$

These facts motivate the following definition.

Definition 2.1. A strategy $\pi_*^1 \in \Pi^1$ is said to be asymptotically discounted optimal (AD-optimal) for player 1 if

$$\liminf_{t \to \infty} E_x^{\pi_*^1, \pi^2} D(x_t, a_t, b_t) \geq 0 \quad \forall x \in X, \ \pi^2 \in \Pi^2.$$

Similarly, $\pi_*^2 \in \Pi^2$ is said to be AD-optimal for player 2 if

$$\limsup_{t \to \infty} E_x^{\pi^1, \pi_*^2} D(x_t, a_t, b_t) \leq 0 \quad \forall x \in X, \ \pi^1 \in \Pi^1.$$

A pair of AD-optimal strategies is called an AD-optimal pair. In this case, if (π_*^1, π_*^2) is an AD-optimal pair, then, for each $x \in X$,

$$\lim_{t \to \infty} E_x^{\pi_*^1, \pi_*^2} D(x_t, a_t, b_t) = 0 \quad \forall x \in X.$$

2.3 Estimation and Control

We consider a discrete-time two person zero-sum Markov game whose *state process* $\{x_t\} \subset X$ evolves according to the equation (see Sect. 1.1.1)

$$x_{t+1} = F(x_t, a_t, b_t, \xi_t), \quad t = 0, 1, \dots, \tag{2.18}$$

where $F : \mathbb{K} \times \mathfrak{R}^k \to X$ is a given measurable function and $(a_t, b_t) \in A(x_t) \times B(x_t)$. The perturbation process $\{\xi_t\}$ is formed by observable i.i.d. $\mathfrak{R}^k$-valued random variables, for some fixed nonnegative integer k, independent of the initial state x_0, with common probability density $\rho(\cdot)$ which is *unknown* to both players. Moreover, we suppose that the realizations $\xi_0, \xi_1, \dots$ of the disturbance process and the states $x_0, x_1, \dots$ are completely observable.

Unlike the standard game model, observe that the solution to the game given in Theorem 2.5 (or Theorem 2.6) is not accessible to the players. Since ρ is unknown, they need to combine statistical density estimation procedures and control processes to gain some insights on the evolution of the game. Hence, in this case our concern is in a game played over an infinite horizon evolving as follows: at each time $t \in \mathbb{N}_0$, the players observe the state of the game $x_t = x \in X$; next, on the record of a sample $\bar{\xi}_t := (\xi_0, \xi_1, \ldots, \xi_{t-1})$ and possibly taking into account the history of the game, players 1 and 2 get estimates $\rho_t^1 = \rho_t^1(\bar{\xi}_t)$ and $\rho_t^2 = \rho_t^2(\bar{\xi}_t)$ of the unknown density ρ, respectively, and adapt independently their strategies to choose actions $a = a_t(\rho_t^1) \in A(x)$ and $b = b_t(\rho_t^2) \in B(x)$, respectively. As a consequence, player 2 pays the amount $r(x, a, b)$ to player 1 and the system visits a new state $x_{t+1} = x' \in X$ according to the evolution law (see (1.4))

$$Q(D|x,a,b) := \int_{\Re^k} 1_D[F(x,a,b,s)]\rho(s)ds \quad D \in \mathscr{B}(X).$$

As in the standard case, the goal of player 1 (player 2, resp.) is to maximize (minimize, resp.) the total expected α-discounted payoff (1.12) (see (2.1)).

Because the discounted criterion depends strongly on the decision selected at the early stages (precisely when the information about the density ρ is deficient) this estimation and control procedure yields in the best case suboptimal strategies in the long term. Therefore the optimality in this case will be studied under the notion of asymptotic optimality given in Definition 2.1.

Observe that the Shapley operator (2.2) takes the form

$$T_{(\alpha,\rho)}v(x) := T_\alpha v(x)$$

$$= \inf_{\varphi^2 \in \mathbb{B}(x)} \sup_{\varphi^1 \in \mathbb{A}(x)} \left[r(x, \varphi^1, \varphi^2) + \alpha \int_{\Re^k} v(F(x, \varphi^1, \varphi^2, s))\rho(s)ds \right].$$

$$(2.19)$$

Now, to show the existence of minimizers/maximizers we need to impose the following conditions.

Assumption 2.8 *(a) For each $s \in \Re^k$, $F(\cdot, \cdot, \cdot, s)$ is continuous.*

(b) Moreover, there exist constants $\lambda_0 \in (0,1)$, $d_0 \geq 0$, $p > 1$, and $M > 0$ such that for all $(x, a, b) \in \mathbb{K}$ it holds that

$$0 \leq r(x,a,b) \leq MW(x), \tag{2.20}$$

$$\int_{\Re^k} W^p[F(x,a,b,s)]\rho(ds) \leq \lambda_0 W^p(x) + d_0. \tag{2.21}$$

Remark 2.4. (a) From Proposition C.3, Assumption 2.8(a) implies that the mapping

$$(x,a,b) \rightarrow \int_{\Re^k} v(F(x,a,b,s))\mu(ds) \tag{2.22}$$

is continuous for each function $v \in \mathbb{C}(X)$ and each probability measure $\mu(\cdot)$ on $\Re^k$, which yields Assumptions 2.1(c) and 2.3(d) hold true. In fact, it is easy to see that Assumption 2.4(c) holds as well. Thus, from Proposition C.1 (b), if $v(\cdot)$ belongs to $\mathbb{L}(X)$, then the mapping (2.22) is in $\mathbb{L}(\mathbb{K})$. On the other hand, by Jensen's inequality, relation (2.21) implies Assumption 2.4(b) with $\beta := \lambda_0^{1/p}$ and $d := d_0^{1/p}$, that is,

$$\int_{\Re^k} W(F(x,a,b,s))\rho(s)ds \le \beta W(x)+d, \tag{2.23}$$

Then, we have that Assumption 2.8 implies Assumption 2.4. Hence, under Assumptions 2.3(a)–(c) and 2.8, the conclusions of Theorem 2.5 are still valid. Similarly, Theorem 2.6 holds under Assumptions 2.1(a), (b) and 2.8.

(b) Furthermore, provided that Assumption 2.8(a) and inequality (2.23) hold, together with either Assumption 2.3(a)–(c) or Assumption 2.1(a)–(b), the facts established in Remark 2.4(a) are valid for any density σ on $\Re^k$ such that

$$\int_{\Re^k} W(F(x,a,b,s))\sigma(s)ds \le \beta W(x)+d. \tag{2.24}$$

Therefore, from Theorem 2.5 (resp. Theorem 2.6) there exists a value $V_\alpha^\sigma \in \mathbb{L}_W$ (resp. $V_\alpha^\sigma \in \mathbb{C}_W$) such that

$$T_{(\alpha,\sigma)}V_\alpha^\sigma(x) = V_\alpha^\sigma(x), \quad x \in X.$$

(c) Additionally, note that Assumption 2.8(b) yields

$$\sup_{t\in\mathbb{N}_0} E_x^{\pi^1,\pi^2}[W^p(x_t)] < \infty \text{ and } \sup_{t\in\mathbb{N}_0} E_x^{\pi^1,\pi^2}[W(x_t)] < \infty \tag{2.25}$$

for each pair $(\pi^1,\pi^2) \in \Pi^1 \times \Pi^2$ and $x \in X$. Indeed, observe that by iterating inequality (2.24) we get

$$E_x^{\pi^1,\pi^2}W(x_t) \le \beta^t W(x) + \frac{1-\beta^t}{1-\beta}d \quad \forall x \in X, \pi^1 \in \Pi^1, \pi^2 \in \Pi^2, t \in \mathbb{N}.$$

Hence,

$$\sup_{t\in\mathbb{N}_0} E_x^{\pi^1,\pi^2}W(x_t) < \infty \quad \forall x \in X, \pi^1 \in \Pi^1, \pi^2 \in \Pi^2. \tag{2.26}$$

Similarly, inequality (2.21) yields

$$\sup_{t\in\mathbb{N}_0} E_x^{\pi^1,\pi^2}W^p(x_t) < \infty \quad \forall x \in X, \pi^1 \in \Pi^1, \pi^2 \in \Pi^2. \tag{2.27}$$

2.3.1 Density Estimation

Denote by L_1 the space of functions defined on $\Re^k$ which are integrable with respect to the Lebesgue measure, and by $||\cdot||_{L_1}$ the corresponding norm. We show the existence of estimators of $\rho \in L_1$ with suitable properties for the construction of asymptotically optimal strategies for both players. To state this precisely we impose the following assumptions on the density.

Assumption 2.9 *There exists a measurable function $\widetilde{\rho}$ on $\Re^k$, such that $\rho(\cdot) \leq \widetilde{\rho}(\cdot)$ a.e. with respect to the Lebesgue measure.*

Assumption 2.10 *The function*

$$\psi(s) := \sup_{(x,a,b)\in\mathbb{K}} \frac{1}{W(x)} W[F(x,a,b,s)], \quad s \in \Re^k, \tag{2.28}$$

is finite and satisfies the integrability condition

$$\bar{\Psi} := \int_{\Re^k} \psi^2(s)\widetilde{\rho}(s)ds < \infty. \tag{2.29}$$

Note that the function ψ in (2.28) is upper semianalytic, thus universally measurable. Hence, the integral in (2.29) is well defined (see [5, Ch. 7, Section 7.6]).

Let $\xi_0, \xi_1, \ldots, \xi_t, \ldots$ be independent realizations of random vectors with density $\rho(\cdot)$, and let $\widehat{\rho}_t^i(s) := \widehat{\rho}_t^i(s; \xi_0, \xi_1, \ldots, \xi_{t-1})$, for $s \in \Re^k$ and $t \in \mathbb{N}$, be density estimators for players $i = 1, 2$, such that

$$E||\widehat{\rho}_t^i - \rho||_{L_1} = E \int_{\Re^k} |\widehat{\rho}_t^i(s) - \rho(s)| \, ds \to 0, \quad \text{as} \quad t \to \infty. \tag{2.30}$$

A wide class of estimators that satisfies this condition are, for instance, the kernel estimates

$$\widehat{\rho}_t^i(s) = \frac{1}{td_t^k} \sum_{i=1}^t K\left(\frac{s - \xi_i}{d_t}\right),$$

where the kernel K is a nonnegative measurable function with $\int_{\Re^k} K(s)ds = 1$, $d_t \to 0$, and $td_t^k \to \infty$ as t goes to infinity (see Appendix D.2.1).

Now, we define the following class of densities.

Definition 2.2. Let $\mathscr{D}$ be the set of densities $\sigma(\cdot)$ on $\Re^k$ satisfying the following conditions:

D.1. $\sigma(\cdot) \leq \widetilde{\rho}(\cdot)$ a.e. with respect to the Lebesgue measure;

D.2. $\int_{\Re^k} W[F(x,a,b,s)]\sigma(s)ds \leq \beta W(x) + d \ \forall (x,a,b) \in \mathbb{K},$

where the function $\widetilde{\rho}$ and the constants β and d are as in Assumption 2.9 and Remark 2.4(a).

Remark 2.5. In the scenario of Assumption 2.7, Condition D.2 is replaced by

$$\mathbf{D.2'}. \quad \int_{\mathfrak{R}^k} W[F(x,a,b,s)]\sigma(s)ds \leq \widetilde{\gamma}W(x) \quad \forall(x,a,b) \in \mathbb{K},$$

where $\widetilde{\gamma} \in \left(1, \dfrac{1}{\alpha}\right).$

Proposition 2.1. *Under Assumption 2.10, the set $\mathscr{D}$ is a closed and convex subset of L_1.*

Proof. First observe that the convexity of $\mathscr{D}$ follows directly from the Definition 2.2. We then proceed to prove that $\mathscr{D}$ is closed. To this end, let us fix a sequence $\{\sigma_n\} \subset \mathscr{D}$ such that

$$\sigma_n \overset{L_1}{\to} \sigma \in L_1 \quad \text{as} \quad n \to \infty. \tag{2.31}$$

Let l be the Lebesgue measure on $\mathfrak{R}^k$ and suppose that there is a set $G \subset \mathfrak{R}^k$ with $l(G) > 0$ and such that $\sigma(s) > \widetilde{\rho}(s)$, $s \in G$. Then, for some $\varepsilon > 0$ and $G' \subset G$ with $l(G') > 0$

$$\sigma(s) > \widetilde{\rho}(s) + \varepsilon, \quad \forall s \in G'. \tag{2.32}$$

On the other hand, since $\sigma_n \in \mathscr{D}$, for $n \in \mathbb{N}_0$, there exists $G_n \subset [0,\infty)$ with $l(G_n) = 0$, such that

$$\sigma_n(s) \leq \widetilde{\rho}(s) \quad \text{for } s \in \mathfrak{R}^k \backslash G_n, \ n \in \mathbb{N}_0. \tag{2.33}$$

Combining (2.32) and (2.33) we get

$$|\sigma_n(s) - \sigma(s)| \geq \varepsilon, \quad \forall s \in G' \cap (\mathfrak{R}^k \backslash G_n), \quad n \in \mathbb{N}_0.$$

Using the fact that $l(G' \cap (\mathfrak{R}^k \backslash G_n)) > 0$ we obtain that σ_n does not converge to σ in measure, which is a contradiction to the convergence in L_1. Therefore $\sigma(s) \leq \widetilde{\rho}(s)$ a.e.

Now we will prove that σ satisfies the inequality in Condition **D.2**. Because

$$\int_{\mathfrak{R}^k} W[F(x,a,b,s)]\sigma_n(s)ds \leq \beta W(x) + d \ \forall(x,a,b) \in \mathbb{K}, n \in \mathbb{N}_0,$$

it is enough to show that for all $(x,a,b) \in \mathbb{K}$,

$$\int_{\mathfrak{R}^k} W[F(x,a,b,s)]\sigma_n(s)ds \to \int_{\mathfrak{R}^k} W[F(x,a,b,s)]\sigma(s)ds, \tag{2.34}$$

as $n \to \infty$. From Assumption 2.10, we have that $W[F(x,a,b,s)] \leq W(x)\psi(s)$ for all $(x,a,b) \in \mathbb{K}$, $s \in \mathfrak{R}^k$. Then

$$
\begin{aligned}
I_n \; : \; = & \left| \int_{\mathfrak{R}^k} W[F(x,a,b,s)]\,[\sigma_n(s) - \sigma(s)]\,ds \right| \\
\leq & \; W(x) \int_{\mathfrak{R}^k} \psi(s)\,|\sigma_n(s) - \sigma(s)|\,ds \\
\leq & \; W(x) \int_{\mathfrak{R}^k} \psi(s)\,|\sigma_n(s) - \sigma(s)|^{\frac{1}{2}}\,|\sigma_n(s) - \sigma(s)|^{\frac{1}{2}}\,ds.
\end{aligned}
$$

By applying Hölder's inequality and taking into account that $\sigma_n(\cdot) \leq \widetilde{\rho}(\cdot)$ and $\sigma(\cdot) \leq \widetilde{\rho}(\cdot)$ a.e., we get

$$
\begin{aligned}
0 \leq I_n \leq \; & W(x) \left[\int_{\mathfrak{R}^k} \psi^2(s)\,|\sigma_n(s) - \sigma(s)| \right]^{1/2} \left[\int_{\mathfrak{R}^k} |\sigma_n(s) - \sigma(s)|\,ds \right]^{1/2} \\
\leq \; & W(x) \left[\int_{\mathfrak{R}^k} \psi^2(s)\,(2\widetilde{\rho}(s)) \right]^{1/2} \left[\int_{\mathfrak{R}^k} |\sigma_n(s) - \sigma(s)|\,ds \right]^{1/2} \\
\leq \; & 2^{1/2} \bar{\Psi}^{1/2} W(x) \left[\int_{\mathfrak{R}^k} |\sigma_n(s) - \sigma(s)|\,ds \right]^{1/2}, \qquad (2.35)
\end{aligned}
$$

where the last inequality comes from (2.29). Hence, letting $n \to \infty$ in (2.35), from (2.31) we get $I_n \to 0$, which yields (2.34).

Finally, we prove that σ is a density. To this end, we first observe that $\sigma(\cdot) \geq 0$ a.e. Now, similar as (2.35),

$$
\begin{aligned}
\left| 1 - \int_{\mathfrak{R}^k} \sigma(s)ds \right| \leq & \int_{\mathfrak{R}^k} |\sigma_n(s) - \sigma(s)|\,ds \\
\leq & \left[\int_{\mathfrak{R}^k} 2\widetilde{\rho}(s) \right]^{1/2} \left[\int_{\mathfrak{R}^k} |\sigma_n(s) - \sigma(s)|\,ds \right]^{1/2} \to 0 \;\; \text{as } n \to \infty.
\end{aligned}
$$

Since $\psi(\cdot) \geq 1$, this fact and (2.29) imply $\int_{\mathfrak{R}^k} \sigma(s)ds = 1$, and therefore σ is a density on $\mathfrak{R}^k$. This proves that D is closed. ∎

From Proposition 2.1 and Theorem D.5 in Appendix D we can use a projection density estimation method to obtain our estimator. That is, for each $t \in \mathbb{N}$, there exists $\rho_t^i(\cdot) \in \mathscr{D}$ such that

$$
||\rho_t^i - \widehat{\rho}_t^i||_{L_1} = \inf_{\sigma \in \mathscr{D}} ||\sigma - \widehat{\rho}_t^i||_{L_1} \;\; \text{for } i = 1,2. \qquad (2.36)
$$

Notice that $E||\rho_t^i - \rho||_{L_1} \to 0$ since (see (2.30))

$$
||\rho_t^i - \rho||_{L_1} \leq ||\rho_t^i - \widehat{\rho}_t^i||_{L_1} + ||\widehat{\rho}_t^i - \rho||_{L_1} \leq 2||\widehat{\rho}_t^i - \rho||_{L_1} \;\; \forall t \in \mathbb{N}. \qquad (2.37)
$$

The densities $\rho_t^i(\cdot)$ defined for $i = 1, 2$, as

$$\rho_t^i := \underset{\sigma \in \mathscr{D}}{\operatorname{argmin}} ||\sigma - \widehat{\rho}_t^i||_{L_1} \qquad (2.38)$$

will be used as estimators to obtain asymptotically optimal strategies for each player. First, however, we will express their convergence property using a more suitable norm defined as follows. For a measurable function $\sigma : \mathfrak{R}^k \to \mathfrak{R}$ define

$$||\sigma|| := \sup_{x \in X} \sup_{a \in A(x), b \in B(x)} \frac{1}{W(x)} \int_{\mathfrak{R}^k} W[F(x,a,b,s)] |\sigma(s)| ds. \qquad (2.39)$$

Note that Condition **D.2** guarantees that $||\sigma|| < \infty$ for any density σ in $\mathscr{D}$.

Lemma 2.1. *If Assumptions 2.9 and 2.10 hold, then*

$$E||\rho_t^i - \rho|| \to 0 \ \text{ as } t \to \infty \ \text{ for } i = 1, 2.$$

Proof. Proceeding similarly as in (2.35), from Holder's inequality, (2.28) and (2.29),

$$||\rho_t^i - \rho|| \le \int_{\mathfrak{R}^k} \psi(s) |\rho_t^i(s) - \rho(s)| ds$$

$$\le \left(\int_{\mathfrak{R}^k} \psi^2(s) |\rho_t^i(s) - \rho(s)| ds \right)^{1/2} \left(\int_{\mathfrak{R}^k} |\rho_t^i(s) - \rho(s)| ds \right)^{1/2}$$

$$\le C' \left(\int_{\mathfrak{R}^k} |\rho_t^i(s) - \rho(s)| ds \right)^{1/2} = C' ||\rho_t^i - \rho||_{L_1}^{1/2}, \qquad (2.40)$$

where $C' := 2^{1/2} \bar{\Psi}^{1/2}$. The result follows from (2.40) since $E||\rho_t^i - \rho||_{L_1} \to 0$ as $t \to \infty$, which, from Remark D.1 (Appendix D), implies $E||\rho_t^i - \rho||_{L_1}^{1/2} \to 0$ as $t \to \infty$. ∎

2.3.2 Asymptotically Optimal Strategies

To obtain asymptotically optimal strategies we will use a nonstationary value iteration approach when the players use the density estimators $\{\rho_t^i\}$ as the true density. Hence, for $i = 1, 2$, we introduce the operators

$$T_{(\alpha, \rho_t^i)} v(x) := \inf_{\varphi^2 \in \mathbb{B}(x)} \sup_{\varphi^1 \in \mathbb{A}(x)} \left[r(x, \varphi^1, \varphi^2) + \alpha \int_{\mathfrak{R}^k} v\left(F(x, \varphi^1, \varphi^2, s)\right) \rho_t^i(s) ds \right]$$

for $x \in X$ and $t \in \mathbb{N}$. Note that these operators are contractions from $(\mathbb{L}_W, || \cdot ||_{\bar{W}})$ into itself with modulus γ_α —see Condition **D.2** and Remark 2.3(a)—provided that Assumptions 2.3(a)–(c) and 2.8 hold. Thus, in this case, from Remark 2.1, the "minimax" value iteration functions for player i ($=1, 2$)

$$U_t^i := T_{(\alpha, \rho_t^i)} U_{t-1}^i \quad t \in \mathbb{N}, \quad U_0^i \equiv 0,$$

belong to the space $\mathbb{L}_W$. Moreover, there exists a sequence $\{\bar{\varphi}_t^2\} \subset \Phi^2$ for player 2 such that

$$U_t^2(x) = \sup_{\varphi^1 \in \mathbb{A}(x)} \left[r(x, \varphi^1, \bar{\varphi}_t^2) + \alpha \int_{\Re^k} U_{t-1}^2 \left(F(x, \varphi^1, \bar{\varphi}_t^2, s) \right) \rho_t^2(s) ds \right] \quad (2.41)$$

for all $x \in X, t \in \mathbb{N}$. Similarly, for any sequence of positive numbers $\{\varepsilon_t\}$ converging to zero there exists a sequence $\{\bar{\varphi}_t^1\} \subset \Phi^1$ such that

$$U_t^1(x) - \varepsilon_t \leq \inf_{\varphi^2 \in \mathbb{B}(x)} \left[r(x, \bar{\varphi}_t^1, \varphi^2) + \alpha \int_{\Re^k} U_{t-1}^1 \left(F(x, \bar{\varphi}_t^1, \varphi^2, s) \right) \rho_t^1(s) ds \right] \quad (2.42)$$

for all $x \in X, t \in \mathbb{N}$. Furthermore, a straightforward calculation shows that, for some positive constants C_1 and C_2,

$$\left| U_t^i(x) \right| \leq C_i W(x) \quad \forall t \in \mathbb{N}_0, \ x \in X, \ i = 1, 2. \quad (2.43)$$

Remark 2.6. In the particular case when $\rho_t^1 = \rho_t^2 = \rho_t$, the sequences $\{\bar{\varphi}_t^1\} \subset \Phi^1$ and $\{\bar{\varphi}_t^2\} \subset \Phi^2$ are defined through the value iteration functions

$$U_t := T_{(\alpha, \rho_t)} U_{t-1} \quad t \in \mathbb{N}, \quad U_0 \equiv 0.$$

That is, for $x \in X$ and $t \in \mathbb{N}$,

$$U_t(x) = \inf_{\varphi^2 \in \mathbb{B}(x)} \sup_{\varphi^1 \in \mathbb{A}(x)} \left[r(x, \varphi^1, \varphi^2) + \alpha \int_{\Re^k} U_{t-1} \left(F(x, \varphi^1, \varphi^2, s) \right) \rho_t(s) ds \right]$$

$$= \sup_{\varphi^1 \in \mathbb{A}(x)} \left[r(x, \varphi^1, \bar{\varphi}_t^2) + \alpha \int_{\Re^k} U_{t-1} \left(F(x, \varphi^1, \bar{\varphi}_t^2, s) \right) \rho_t(s) ds \right]$$

$$\leq \inf_{\varphi^2 \in \mathbb{B}(x)} \left[r(x, \bar{\varphi}_t^1, \varphi^2) + \alpha \int_{\Re^k} U_{t-1} \left(F(x, \bar{\varphi}_t^1, \varphi^2, s) \right) \rho_t(s) ds \right] + \varepsilon_t. \quad (2.44)$$

In the scenario of Assumptions 2.1 (or 2.2) and 2.4, we can replace (2.44) by the following equation

$$U_t(x) = \inf_{\varphi^2 \in \mathbb{B}(x)} \sup_{\varphi^1 \in \mathbb{A}(x)} \left[r(x, \varphi^1, \varphi^2) + \alpha \int_{\mathfrak{R}^k} U_{t-1}\left(F(x, \varphi^1, \varphi^2, s)\right) \rho_t(s) ds \right]$$

$$= \sup_{\varphi^1 \in \mathbb{A}(x)} \left[r(x, \varphi^1, \bar{\varphi}_t^2) + \alpha \int_{\mathfrak{R}^k} U_{t-1}\left(F(x, \varphi^1, \bar{\varphi}_t^2, s)\right) \rho_t(s) ds \right]$$

$$= \inf_{\varphi^2 \in \mathbb{B}(x)} \left[r(x, \bar{\varphi}_t^1, \varphi^2) + \alpha \int_{\mathfrak{R}^k} U_{t-1}\left(F(x, \bar{\varphi}_t^1, \varphi^2, s)\right) \rho_t(s) ds \right]. \qquad (2.45)$$

Finally we state the main result as follows.

Theorem 2.11. *Suppose that Assumptions 2.3(a)–(c), 2.8, 2.9, and 2.10 hold. Then the strategies $\bar{\pi}_*^1 = \{\bar{\varphi}_t^1\} \in \Pi^1$ and $\bar{\pi}_*^2 = \{\bar{\varphi}_t^2\} \in \Pi^2$ are AD-optimal for player 1 and player 2, respectively. Thus, in particular,*

$$\lim_{t \to \infty} E_x^{\bar{\pi}_*^1, \bar{\pi}_*^2} D(x_t, a_t, b_t) = 0 \quad \forall x \in X.$$

2.3.3 Proof of Theorem 2.11

The proof of Theorem 2.11 is a consequence of Lemmas 2.2–2.5 below. Throughout this subsection we suppose that all the assumptions in Theorem 2.11 hold.

Lemma 2.2. *For each i=1, 2:*

$$\left\| T_{(\alpha, \rho_t^i)} V_\alpha^\rho - T_{(\alpha, \rho)} V_\alpha^\rho \right\|_W \le \alpha \left\| V_\alpha^\rho \right\|_W \left\| \rho_t^i - \rho \right\| \quad \forall t \in \mathbb{N}. \qquad (2.46)$$

Thus, for each $(\pi^1, \pi^2) \in \Pi^1 \times \Pi^2$ and $x \in X$,

$$\lim_{t \to \infty} E_x^{\pi^1, \pi^2} \left\| T_{(\alpha, \rho_t^i)} V_\alpha^\rho - T_{(\alpha, \rho)} V_\alpha^\rho \right\|_W = 0. \qquad (2.47)$$

Proof. Because $\left\| \rho_t^i - \rho \right\|$ does not depend on $(\pi^1, \pi^2) \in \Pi^1 \times \Pi^2$ and $x \in X$, note that (2.47) follows from (2.46) and Lemma 2.1. Thus we proceed to prove (2.46). To do this, for all $x \in X$, $t \in \mathbb{N}$, and $i = 1, 2$, we have

$$\left| T_{(\alpha, \rho_t^i)} V_\alpha^\rho(x) - T_{(\alpha, \rho)} V_\alpha^\rho(x) \right|$$

$$\le \sup_{\varphi^2 \in \mathbb{B}(x)} \sup_{\varphi^1 \in \mathbb{A}(x)} \left| \alpha \int_{\mathfrak{R}^k} V_\alpha^\rho\left(F(x, \varphi^1, \varphi^2, s)\right) \rho_t^i(s) ds \right.$$

$$\left. - \alpha \int_{\mathfrak{R}^k} V_\alpha^\rho\left(F(x, \varphi^1, \varphi^2, s)\right) \rho(s) ds \right|$$

$$\leq \sup_{\varphi^2 \in \mathbb{B}(x)} \sup_{\varphi^1 \in \mathbb{A}(x)} \alpha \int_{\Re^k} \left| V_\alpha^\rho \left(F(x, \varphi^1, \varphi^2, s) \right) \right| \left| \rho_t^i(s) - \rho(s) \right| ds$$

$$\leq \alpha \left\| V_\alpha^\rho \right\|_W \sup_{\varphi^2 \in \mathbb{B}(x)} \sup_{\varphi^1 \in \mathbb{A}(x)} \int_{\Re^k} W \left(F(x, \varphi^1, \varphi^2, s) \right) \left| \rho_t^i(s) - \rho(s) \right| ds$$

$$\leq \alpha \left\| V_\alpha^\rho \right\|_W \left\| \rho_t^i - \rho \right\| W(x).$$

Hence, (2.46) holds. ∎

Lemma 2.3. *The following holds:*

$$\lim_{t \to \infty} E_x^{\pi^1, \pi^2} \left\| U_t^i - V_\alpha^\rho \right\|_W = 0$$

for each $i = 1, 2$, $(\pi^1, \pi^2) \in \Pi^1 \times \Pi^2$, *and* $x \in X$.

Proof. First note that for all $t \in \mathbb{N}$ and $i = 1, 2$,

$$\left\| U_t^i - V_\alpha^\rho \right\|_{\bar{W}} = \left\| T_{(\alpha, \rho_t^i)} U_{t-1}^i - T_{(\alpha, \rho)} V_\alpha^\rho \right\|_{\bar{W}}$$

$$\leq \left\| T_{(\alpha, \rho_t^i)} U_{t-1}^i - T_{(\alpha, \rho_t^i)} V_\alpha^\rho \right\|_{\bar{W}} + \left\| T_{(\alpha, \rho_t^i)} V_\alpha^\rho - T_{(\alpha, \rho)} V_\alpha^\rho \right\|_{\bar{W}}.$$

Then, since $T_{(\alpha, \rho_t^i)}$ is a contraction operator with modulus γ_α with respect to the $\bar{W}$-norm, it follows from Lemma 2.2 that

$$\left\| U_t^i - V_\alpha^\rho \right\|_{\bar{W}} \leq \gamma_\alpha \left\| U_{t-1}^i - V_\alpha^\rho \right\|_{\bar{W}} + \alpha \left\| V_\alpha^\rho \right\|_W \left\| \rho_t^i - \rho \right\|. \tag{2.48}$$

Moreover, since $E \left\| \rho_t^i - \rho \right\| \to 0$, there exists a positive constant M' such that

$$E_x^{\pi^1, \pi^2} \left\| U_t^i - V_\alpha^\rho \right\|_{\bar{W}} \leq \gamma_\alpha E_x^{\pi^1, \pi^2} \left\| U_{t-1}^i - V_\alpha^\rho \right\|_{\bar{W}} + M'.$$

Thus, by iterations of this inequality we obtain

$$E_x^{\pi^1, \pi^2} \left\| U_t^i - V_\alpha^\rho \right\|_{\bar{W}} \leq (\gamma_\alpha)^t \left\| V_\alpha^\rho \right\|_{\bar{W}} + M' \sum_{k=0}^{t-1} (\gamma_\alpha)^k$$

$$\leq \frac{M' + \left\| V_\alpha^\rho \right\|_{\bar{W}}}{1 - \gamma_\alpha},$$

which in turn implies that

$$l := \limsup_{t \to \infty} E_x^{\pi^1, \pi^2} \left\| U_t^i - V_\alpha^\rho \right\|_{\bar{W}} < \infty.$$

Now taking expectation on (2.48) and then limsup as t goes to infinity, we have that $0 \leq l \leq \gamma_\alpha l$, which yields that $l = 0$. This proves the desired result. ∎

In the following we shall use the "approximate discrepancy functions" for player i (=1,2) defined as

$$D_t^i(x, \varphi^1, \varphi^2) := r(x, \varphi^1, \varphi^2) + \alpha \int_{\Re^k} U_{t-1}^i(F(x, \varphi^1, \varphi^2, s)) \rho_t^i(s) ds - U_t^i(x)$$

for $x \in X, \varphi^1 \in \mathbb{A}(x), \varphi^2 \in \mathbb{B}(x)$.

Lemma 2.4. *For all $x \in X$ and $t \in \mathbb{N}$, it holds that*

$$\sup_{\varphi^1 \in \mathbb{A}(x)} \sup_{\varphi^2 \in \mathbb{B}(x)} |D(x, \varphi^1, \varphi^2) - D_t^i(x, \varphi^1, \varphi^2)| \leq W(x) \eta_t^i,$$

where

$$\eta_t^i := \|U_t^i - V_\alpha^\rho\|_W + (\beta + d) \|U_{t-1}^i - V_\alpha^\rho\|_W + \alpha \|V_\alpha^\rho\|_W \|\rho_t^i - \rho\| \qquad (2.49)$$

for all $t \in \mathbb{N}$.

Proof. Let $x \in X, \varphi^1 \in \mathbb{A}(x), \varphi^2 \in \mathbb{B}(x)$, and $t \in \mathbb{N}$ be fixed but arbitrary, and write $R_t^i(x, \varphi^1, \varphi^2) := |D(x, \varphi^1, \varphi^2) - D_t^i(x, \varphi^1, \varphi^2)|$. Then, observe that

$$R_t^i(x, \varphi^1, \varphi^2) \leq |U_t^i(x) - V_\alpha^\rho(x)|$$

$$+ \alpha \left| \int_{\Re^k} U_{t-1}^i(F(x, \varphi^1, \varphi^2, s)) \rho_t^i(s) - \int_{\Re^k} V_\alpha^\rho((F(x, \varphi^1, \varphi^2, s)) \rho(s) \right|$$

$$\leq |U_t^i(x) - V_\alpha^\rho(x)| + \alpha \int_{\Re^k} V_\alpha^\rho((F(x, \varphi^1, \varphi^2, s)) |\rho(s) - \rho_t^i(s)| ds$$

$$+ \alpha \int_{\Re^k} |U_{t-1}^i(F(x, \varphi^1, \varphi^2, s)) - V_\alpha^\rho(F(x, \varphi^1, \varphi^2, s))| \rho_t^i(s) ds$$

$$\leq |U_t^i(x) - V_\alpha^\rho(x)| + \alpha \|V_\alpha^\rho\|_W \|\rho_t^i - \rho\| W(x)$$

$$+ \|U_{t-1}^i - V_\alpha^\rho\|_W \int_{\Re^k} W(F(x, \varphi^1, \varphi^2, s)) \rho_t^i(s) ds.$$

Now, since $\rho_t^i(\cdot)$ is in $\mathscr{D}$, from Condition **D.2** we have

$$\int_{\Re^k} W[F(x,a,b,s)]\rho_t^i(s)ds \le \beta W(x)+d$$

$$\le [\beta+d]W(x).$$

This implies

$$R_t^i(x,\varphi^1,\varphi^2) \le \left|U_t^i(x)-V_\alpha^\rho(x)\right|+\alpha\left\|V_\alpha^\rho\right\|_W\left\|\rho_t^i-\rho\right\|_W W(x)$$

$$+\left\|U_{t-1}^i-V_\alpha^\rho\right\|_W (\beta+d)W(x).$$

Hence,

$$\sup_{\varphi^1\in\mathbb{A}(x)}\sup_{\varphi^2\in\mathbb{B}(x)}\left|D(x,\varphi^1,\varphi^2)-D_t^i(x,\varphi^1,\varphi^2)\right|\le W(x)\eta_t^i$$

for all $x\in X,t\in\mathbb{N}$. ∎

Lemma 2.5. *Let the strategies $\pi_*^1=\{\bar{\varphi}_t^1\}$ and $\pi_*^2=\{\bar{\varphi}_t^2\}$ be as in (2.42) and (2.41). Then*

$$-\varepsilon_t-W(x_t)\eta_t^1\le D(x,\bar{\varphi}_t^1,\varphi^2)\quad\forall\varphi^2\in\mathbb{B}(x),t\in\mathbb{N},$$

$$D(x,\varphi^1,\bar{\varphi}_t^2)\le W(x_t)\eta_t^2\quad\forall\varphi^1\in\mathbb{A}(x),t\in\mathbb{N},$$

with η_t^i as in (2.49).

Proof. The inequalities follow directly from Lemma 2.4 noting that

$$\inf_{\varphi^2\in\mathbb{B}(x)}D_t^1(x,\bar{\varphi}_t^1,\varphi^2)\ge-\varepsilon_t,$$

and

$$\sup_{\varphi^1\in\mathbb{A}(x)}D_t^2(x,\varphi^1,\bar{\varphi}_t^2)=0$$

hold for all $x\in X,t\in\mathbb{N}$. ∎

Remark 2.7. Observe that from (2.49), and Lemmas 2.1 and 2.3,

$$\lim_{t\to\infty}E_x^{\pi^1,\pi^2}\eta_t^i=0\quad\text{for }i=1,2,\ (\pi^1,\pi^2)\in\Pi^1\times\Pi^2,\text{ and }x\in X.$$

Thus, for $i=1,2$,

$$\eta_t^i\to0\text{ in probability }P_x^{\pi^1,\pi^2}.$$

Moreover, since $||\sigma|| < \infty$ for σ in $\mathscr{D}$, from (2.43) we have

$$k_i := \sup_t \eta_t^i < \infty \quad \text{for } i = 1, 2.$$

Finally we are ready to prove our main theorem in this chapter.

Proof (Proof of Theorem 2.11). From Lemma 2.5 it is enough to prove, for $i=1, 2$, that

$$E_x^{\pi^1,\pi^2} W(x_t)\eta_t^i \to 0 \quad \forall x \in X, \pi^1 \in \Pi^1, \pi^2 \in \Pi^2. \tag{2.50}$$

To prove this fact, we begin by proving that the process $\{W(x_t)\eta_t^i\}$ converges to zero in probability with respect to $P_x^{\pi^1,\pi^2}$ for all $x \in X, \pi^1 \in \Pi^1, \pi^2 \in \Pi^2$. To do this, let l_1 and l_2 be arbitrary positive constants; then observe that for all $x \in X$ and $t \in \mathbb{N}$,

$$P_x^{\pi^1,\pi^2}\left[W(x_t)\eta_t^i > l_1\right] \leq P_x^{\pi^1,\pi^2}\left[\eta_t^i > \frac{l_1}{l_2}\right] + P_x^{\pi^1,\pi^2}\left[W(x_t) > l_2\right].$$

Thus, Chebyshev's inequality and (2.25) yield

$$P_x^{\pi^1,\pi^2}\left[W(x_t)\eta_t^i > l_1\right] \leq P_x^{\pi^1,\pi^2}\left[\eta_t^i > \frac{l_1}{l_2}\right] + \frac{1}{l_2}E_x^{\pi^1,\pi^2}W(x_t)$$

$$\leq P_x^{\pi^1,\pi^2}\left[\eta_t^i > \frac{l_1}{l_2}\right] + \frac{\bar{M}}{l_2}$$

for some constant $\bar{M} < \infty$. Hence, from Remark 2.7,

$$\limsup_{t\to\infty} P_x^{\pi^1,\pi^2}\left[W(x_t)\eta_t^i > l_1\right] \leq \frac{\bar{M}}{l_2}.$$

Since l_2 is arbitrary, we conclude that

$$\lim_{t\to\infty} P_x^{\pi^1,\pi^2}\left[W(x_t)\eta_t^i > l_1\right] = 0,$$

which proves the claim.

On the other hand, from (2.25) and Remark 2.7, we see that the inequality

$$\sup_{t\in\mathbb{N}} E_x^{\pi^1,\pi^2}\left[W(x_t)\eta_t^i\right]^p \leq k_i^p \sup_{t\in\mathbb{N}} E_x^{\pi^1,\pi^2} W^p(x_t) < \infty$$

holds for all $x \in X, \pi^1 \in \Pi^1, \pi^2 \in \Pi^2$. Thus, from [3, Lemma 7.6.9, p. 301], the latter inequality implies that the process $\{W(x_t)\eta_t^i\}$ is $P_x^{\pi^1,\pi^2}$-uniformly integrable.

Finally, using the uniform integrability of the process $\{W(x_t)\eta_t^i\}$ and that it converges to zero, we conclude that

$$E_x^{\pi^1,\pi^2} W(x_t)\eta_t^i \to 0$$

as $t \to \infty$. ∎

Remark 2.8 (Bounded Payoffs). The idea behind the use of projection estimates is that we need an estimator that satisfies Condition **D.2**. As is noted in Remark 2.2, if the payoff function r is bounded we can take $W \equiv 1$, so Assumption 2.10 holds letting $\rho = \bar{\rho}$, and we have $\|\cdot\|_{L_1} = \|\cdot\|$ (see (2.39)). Thus, any L_1-consistent density estimates ρ_t^i, $i = 1, 2$, can be used for the construction of strategies under estimation and control procedures. Moreover, from (A.1) in Appendix A, the norms $\|\cdot\|_B$ and $\|\cdot\|_W$ coincide, which yields that the convergences stated in Lemmas 2.2 and 2.3 would be under the norm $\|\cdot\|_B$.

Chapter 3
Average Payoff Criterion

We are now interested in analyzing the average payoff game, that is, the game in (1.3) with the long-run expected average payoff in (1.13):

$$J(x, \pi^1, \pi^2) := \liminf_{n \to \infty} \frac{1}{n} E_x^{\pi^1, \pi^2} \sum_{t=0}^{n-1} r(x_t, a_t, b_t) \tag{3.1}$$

for $(\pi^1, \pi^2) \in \Pi^1 \times \Pi^2$ and $x \in X$. Our first concern is to establish conditions ensuring the existence of a value of the game, i.e., a function $J(\cdot)$ such that $U(x) = L(x) = J(x)$, where U and L are the upper and the lower values of the average game (see (1.17)–(1.19)). Having the value of the game, we then give conditions ensuring the existence of a saddle point $(\pi_*^1, \pi_*^2) \in \Pi^1 \times \Pi^2$, that is,

$$J(x, \pi^1, \pi_*^2) \le J(x, \pi_*^1, \pi_*^2) = J(x) \le J(x, \pi_*^1, \pi^2) \ \forall x \in X, (\pi^1, \pi^2) \in \Pi^1 \times \Pi^2.$$

Finally, as in the discounted game, we introduce estimation and control procedures in difference equation game models (see Sect. 1.1.1), assuming that $\{\xi_t\}$ is a sequence of i.i.d. r.v. with unknown density ρ.

Unlike the discounted criterion (1.12), the average performance index depends on the tail of the game's state process instead of the early stages of the game. Hence, an asymptotic analysis is necessary to study the average payoff game, so suitable ergodicity conditions should be imposed on the game model. Due precisely to this asymptotic analysis, estimation and control procedures do provide average optimal strategies.

As in the field of Markov Decision Processes (MDPs), the average criterion in games can be studied by means of the so-called *vanishing discount factor approach* (VDFA), that is, as a limit of the discounted criterion (see [42], for instance). In this book, we apply a variant of the VDFA to obtain average optimal pairs of strategies under estimation and control procedures. Specifically, for a suitable discount factor

© The Author(s), under exclusive license to Springer Nature Switzerland AG 2020
J. A. Minjárez-Sosa, *Zero-Sum Discrete-Time Markov Games
with Unknown Disturbance Distribution*, SpringerBriefs in Probability
and Mathematical Statistics, https://doi.org/10.1007/978-3-030-35720-7_3

sequence $\alpha_t \uparrow 1$ and by replacing the unknown density ρ by the estimates ρ_t^1 and ρ_t^2, obtained in each approximation stage, we can obtain minimizers/maximizers in the corresponding α_t-discounted Shapley equations. Then, with the ergodicity conditions, we analyze the limit as $t \to \infty$.

3.1 Continuity and Ergodicity Conditions

The analysis of the average game will be made under two sets of assumptions. The first one, Assumption 3.1, is a combination of some conditions in Assumptions 2.1 and 2.4, which we rewrite for easy reference. It contains continuity and compactness conditions yielding the existence of measurable selectors. In addition, Assumption 3.2 is an ergodicity condition needed to analyze the asymptotic behavior for the average criterion.

Assumption 3.1 *(a) The multifunctions $x \to A(x)$ and $x \to B(x)$ are compact-valued and continuous.*

(b) The payoff function $r(x,a,b)$ is continuous in $(x,a,b) \in \mathbb{K}$.

(c) The mapping

$$(x,a,b) \to \int_X v(y) Q(dy|x,a,b)$$

is continuous on $\mathbb{K}$, for every function $v \in \mathbb{C}(X)$.

(d) There exist a continuous function $W \geq 1$ on X and a constant $M > 0$ such that $0 \leq r(x,a,b) \leq MW(x)$ for all $(x,a,b) \in \mathbb{K}$.

(e) The mapping $(x,a,b) \to \int_X W(y) Q(dy|x,a,b)$ is continuous on $\mathbb{K}$.

Assumption 3.2 *There exist a measurable function $\lambda : \mathbb{K} \to [0,1]$, a probability measure m^* on X, and a constant $\beta \in (0,1)$ such that:*

(a) $\int_X W(y) Q(dy|x,a,b) \leq \beta W(x) + \lambda(x,a,b)d$ for all $(x,a,b) \in \mathbb{K}$, where

$$d := \int_X W(x) m^*(dx) < \infty.$$

(b) $Q(D|x,a,b) \geq \lambda(x,a,b) m^*(D)$ for all $D \in \mathscr{B}(X), (x,a,b) \in \mathbb{K}$;

(c) $\int_X \bar{\Lambda}(x) m^*(dx) > 0$, where $\bar{\Lambda}(x) := \inf\limits_{a \in A(x)} \inf\limits_{b \in B(x)} \lambda(x,a,b)$ is assumed to be a measurable function.

Assumption 3.2 implies that the game's state processes $\{x_t\}$ defined by the class of stationary strategies are W-geometrically ergodic Markov chains with uniform convergence rate. That is, there exist constants $R > 0$ and $\gamma \in (0,1)$ such that

$$\left| E_x^{\varphi^1, \varphi^2} u(x_t) - \int_X u(y) d\mu_{\varphi^1, \varphi^2}(y) \right| \leq RW(x) \gamma^t \tag{3.2}$$

for all $t \in \mathbb{N}$, $x \in X$, $u \in \mathbb{B}_W$, and strategies $(\varphi^1, \varphi^2) \in \Phi^1 \times \Phi^2$, where $\mu_{\varphi^1, \varphi^2}$ stands for the unique invariant probability measure corresponding to the Markov chain $\{x_t\}$ induced by the pair (φ^1, φ^2). It is important to note that the constants R and γ depend neither on the strategies φ^1 and φ^2 nor on the kernel Q (see, e.g., [23, 24]). These conditions, or some variants of them, have been used in several works ([23, 24, 31, 39, 42, 44, 51, 73]). The reader is referred to these works for detailed discussions about their meanings and consequences.

Remark 3.1. Observe that Assumption 3.2(a) implies the inequality in Assumption 2.4(b). Then, under Assumptions 3.1 and 3.2(a), the conclusions of Theorem 2.6 hold. That is, for each $\alpha \in (0, 1)$:

(a) The discounted game has a value $V_\alpha \in \mathbb{C}_W$ and

$$\|V_\alpha\|_W \leq \frac{M}{1 - \alpha}. \tag{3.3}$$

(b) The value V_α satisfies

$$T_\alpha V_\alpha = V_\alpha, \tag{3.4}$$

and there exists $(\varphi_*^1, \varphi_*^2) \in \Phi^1 \times \Phi^2$, such that $\varphi_*^1(x) \in \mathbb{A}(x)$ and $\varphi_*^2 \in \mathbb{B}(x)$ satisfy

$$V_\alpha(x) = r(x, \varphi_*^1, \varphi_*^2) + \alpha \int_X V_\alpha(y) Q(dy | x, \varphi_*^1, \varphi_*^2)$$

$$= \max_{\varphi^1 \in \mathbb{A}(x)} \left[r(x, \varphi^1, \varphi_*^2) + \alpha \int_X V_\alpha(y) Q(dy | x, \varphi^1, \varphi_*^2) \right]$$

$$= \min_{\varphi^2 \in \mathbb{B}(x)} \left[r(x, \varphi_*^1, \varphi^2) + \alpha \int_X V_\alpha(y) Q(dy | x, \varphi_*^1, \varphi^2) \right], \quad \forall x \in X. \tag{3.5}$$

In addition, $\pi_*^1 = \{\varphi_*^1\} \in \Pi_s^1$ and $\pi_*^2 = \{\varphi_*^2\} \in \Pi_s^2$ form an α-optimal pair of strategies.

3.2 The Vanishing Discount Factor Approach (VDFA)

Suppose that Assumptions 3.1 and 3.2 hold. Let V_α and $(\{\varphi_*^1\}, \{\varphi_*^2\}) \in \Pi_s^1 \times \Pi_s^2$ be the value of the discounted game and a pair of discounted optimal strategies (see Remark 3.1(b)). Let $z \in X$ be an arbitrary fixed state and define

$$h_\alpha(\cdot) := V_\alpha(\cdot) - V_\alpha(z) \quad \text{and} \quad j_\alpha := (1 - \alpha) V_\alpha(z)$$

for any $\alpha \in (0,1)$. Then Eq. (3.4) is equivalent to

$$j_\alpha + h_\alpha(x) = T_\alpha h_\alpha(x), \quad x \in X. \tag{3.6}$$

Moreover, from (3.5),

$$
\begin{aligned}
j_\alpha + h_\alpha(x) &= r(x, \varphi_*^1, \varphi_*^2) + \alpha \int_X h_\alpha(y) Q(dy|x, \varphi_*^1, \varphi_*^2) \\
&= \max_{\varphi^1 \in \mathbb{A}(x)} \left[r(x, \varphi^1, \varphi_*^2) + \alpha \int_X h_\alpha(y)(x) Q(dy|x, \varphi^1, \varphi_*^2) \right] \\
&= \min_{\varphi^2 \in \mathbb{B}(x)} \left[r(x, \varphi_*^1, \varphi^2) + \alpha \int_X h_\alpha(y) Q(dy|x, \varphi_*^1, \varphi^2) \right], \quad \forall x \in X.
\end{aligned}
$$

The following theorem, borrowed from [42, Theorem 4.3], states the existence of a value for the average game defined as the limit of j_α as $\alpha \uparrow 1$.

Theorem 3.3. *Suppose Assumptions 3.1 and 3.2 hold. Then the average payoff game has a constant value $J(\cdot) = j^*$, that is,*

$$j^* = \inf_{\pi^2 \in \Pi^2} \sup_{\pi^1 \in \Pi^1} J(x, \pi^1, \pi^2) = \sup_{\pi^1 \in \Pi^1} \inf_{\pi^2 \in \Pi^2} J(x, \pi^1, \pi^2) \quad \forall x \in X.$$

Moreover, there exists an average optimal pair of strategies $(\pi_*^1, \pi_*^2) \in \Pi_s^1 \times \Pi_s^2$ and

$$j^* = \lim_{\alpha \to 1^-} (1 - \alpha) V_\alpha(z), \tag{3.7}$$

where $z \in X$ is an arbitrary fixed state.

The proof essentially is based on the analysis of the limit, as $t \to \infty$, of the α_t-Shapley equation

$$j_{\alpha_t} + h_{\alpha_t}(x) = T_{\alpha_t} h_{\alpha_t}(x), \quad x \in X, \tag{3.8}$$

for a fixed and arbitrary sequence $\{\alpha_t\}$ of discount factors converging to 1. In fact, from (3.7) we have

$$j^* = \lim_{t \to \infty} j_{\alpha_t}. \tag{3.9}$$

3.2.1 Difference Equation Average Game Models

We again consider a game evolving according to a difference equation as in (1.3) (see Sect. 1.1.1), that is,

$$x_{t+1} = F(x_t, a_t, b_t, \xi_t), \quad t = 0, 1, \dots,$$

where $F : \mathbb{K} \times \mathfrak{R}^k \to X$ is a given measurable function, $(a_t, b_t) \in A(x_t) \times B(x_t)$, and $\{\xi_t\}$ is a sequence of i.i.d. $\mathfrak{R}^k$-valued r.v. with density $\rho(\cdot)$.

We assume that for each $s \in \mathfrak{R}^k$ the function $F(\cdot, \cdot, \cdot, s)$ is continuous and there exists a measurable function $\widetilde{\rho} : \mathfrak{R}^k \to \mathfrak{R}^+$ such that $\rho(\cdot) \leq \widetilde{\rho}(\cdot)$ with respect to the

Lebesgue measure (see Assumptions 2.8 and 2.9). We again consider the class $\mathscr{D}$ of densities $\sigma(\cdot)$ satisfying the following conditions (see Definition 2.2):

D.1 $\sigma(\cdot) \leq \widetilde{\rho}(\cdot)$ a.e. with respect to the Lebesgue measure.
D.2   For all $(x, a, b) \in \mathbb{K}$,

$$\int_{\mathfrak{R}^k} W[F(x,a,b,s)]\sigma(s)ds \leq \beta W(x) + d. \tag{3.10}$$

Note that Assumption 3.2(a) implies that the density $\rho(\cdot)$ belongs to $\mathscr{D}$, after observing that the transition kernel Q takes the form (see (1.4))

$$Q(D|x,a,b) = \int_{\mathfrak{R}^k} 1_D[F(x,a,b,s)]\rho(s)ds$$

for $D \in \mathscr{B}(X)$ and $(x,a,b) \in \mathbb{K}$.

Remark 3.2 (cf. Remark 2.4(b)). Under the continuity condition of the function F and Assumptions 3.1(a), (b), (d), we have that for any density σ satisfying (3.10) and $\alpha \in (0,1)$, there exists a function $V_\alpha^\sigma \in \mathbb{C}_W$ which is the value of the corresponding discounted game.

We denote, for a fixed $z \in X$, $\alpha \in (0,1)$, and $\sigma \in \mathscr{D}$

$$h_\alpha^\sigma(\cdot) := V_\alpha^\sigma(\cdot) - V_\alpha^\sigma(z) \quad \text{and} \quad j_\alpha^\sigma := (1-\alpha)V_\alpha^\sigma(z).$$

Observe that the α-discounted Shapley equation (3.6) takes the form

$$j_\alpha^\sigma + h_\alpha^\sigma(x) = \max_{\varphi^1 \in \mathbb{A}(x)} \min_{\varphi^2 \in \mathbb{B}(x)} \left[r(x,\varphi^1,\varphi^2) + \alpha \int_{\mathfrak{R}^k} h_\alpha^\sigma(F(x,\varphi^1,\varphi^2,s))\rho(s)ds \right]. \tag{3.11}$$

Moreover, under Assumption 3.1, for any density $\sigma \in \mathscr{D}$ and $\alpha \in (0,1)$ the following holds (see Remarks 3.1 and 3.2): there exists $(\varphi_*^1, \varphi_*^2) \in \Phi^1 \times \Phi^2$, such that for all $x \in X$, $\varphi_*^1(x) \in \mathbb{A}(x)$ and $\varphi_*^2(x) \in \mathbb{B}(x)$ is satisfied

$$\begin{aligned}
V_\alpha^\sigma(x) &= \max_{\varphi^1 \in \mathbb{A}(x)} \min_{\varphi^2 \in \mathbb{B}(x)} \left[r(x,\varphi^1,\varphi^2) + \alpha \int_X V_\alpha^\sigma(F(x,\varphi^1,\varphi^2,s))\sigma(s)ds \right] \\
&= r(x,\varphi_*^1,\varphi_*^2) + \alpha \int_X V_\alpha^\sigma(F(x,\varphi_*^1,\varphi_*^2,s))\sigma(s)ds \\
&= \max_{\varphi^1 \in \mathbb{A}(x)} \left[r(x,\varphi^1,\varphi_*^2) + \alpha \int_X V_\alpha^\sigma(F(x,\varphi^1,\varphi_*^2,s))\sigma(s)ds \right] \\
&= \min_{\varphi^2 \in \mathbb{B}(x)} \left[r(x,\varphi_*^1,\varphi^2) + \alpha \int_X V_\alpha^\sigma(F(x,\varphi_*^1,\varphi^2,s))\sigma(s)ds \right];
\end{aligned} \tag{3.12}$$

or, equivalently (see (3.6)),

$$j_\alpha^\sigma + h_\alpha^\sigma(x) = r(x, \varphi_*^1, \varphi_*^2) + \alpha \int_X h_\alpha^\sigma(F(x, \varphi_*^1, \varphi_*^2, s)) \sigma(s) ds$$

$$= \max_{\varphi^1 \in \mathbb{A}(x)} \left[r(x, \varphi^1, \varphi_*^2) + \alpha \int_X h_\alpha^\sigma(F(x, \varphi^1, \varphi_*^2, s)) \sigma(s) ds \right]$$

$$= \min_{\varphi^2 \in \mathbb{B}(x)} \left[r(x, \varphi_*^1, \varphi^2) + \alpha \int_X h_\alpha^\sigma(F(x, \varphi_*^1, \varphi^2, s)) \sigma(s) ds \right]. \quad (3.13)$$

3.3 Estimation and Control Under the VDFA

We now introduce estimation and control procedures for the game models described in the previous section. To this end, the disturbance process $\{\xi_t\}$ is assumed to be completely observable. Our approach consists in a combination of the VDFA and suitable density estimation methods. More precisely, to obtain an average optimal pair of strategies, we analyze the limit of the $\hat{\alpha}_t$-discounted Shapley equations (3.11) for a convenient and fixed discount factor sequence $\hat{\alpha}_t \uparrow 1$, and replace the unknown density ρ by the estimates ρ_t^1 and ρ_t^2 obtained in each approximation stage by the players.

We shall show the existence of an average optimal pair of strategies under the following strengthened version of Assumption 3.2 (see Assumptions 2.8(b) and 2.10).

Assumption 3.4 (a) There exist constants $\lambda_0 \in (0,1)$, $d_0 \geq 0$, and $p > 1$ such that

$$\int_{\Re^k} W^p[F(x,a,b,s)] \rho(s) ds \leq \lambda_0 W^p(x) + d_0 \quad (x,a,b) \in \mathbb{K}. \quad (3.14)$$

(b) The function

$$\psi(s) := \sup_{(x,a,b) \in \mathbb{K}} \frac{1}{W(x)} W[F(x,a,b,s)], \quad s \in \Re^k$$

is finite and satisfies the condition

$$\int_{\Re^k} \psi^2(s) (\widetilde{\rho}(s)) ds < \infty.$$

The key point to define the average optimal pairs of strategies is to get density estimators ρ_t^i, $i \in \{1,2\}$, for players 1 and 2 respectively, belonging to the class $\mathscr{D}$, with a property of convergence stronger than that given in Lemma 2.1. As in discounted case, these estimators are obtained by means of a *projection estimate method* with a slight modification (see Appendix D.2.2). Essentially, the requirement that $\rho_t^i \in \mathscr{D}$ is to be able to use the estimated versions of the Shapley equations (3.12) and (3.13).

Let $\xi_0, \xi_1, \ldots, \xi_t, \ldots$ be independent realizations of random variables in $\mathfrak{R}^k$ with density $\rho(\cdot)$, and $\widehat{\rho}_t^i(\cdot) := \widehat{\rho}_t^i(\cdot; \xi_0, \xi_1, \ldots, \xi_{t-1})$, for $t \in \mathbb{N}$, be density estimators for player $i \in \{1, 2\}$ satisfying the property

$$E\|\widehat{\rho}_t^i - \rho\|_{L_1} = E \int_{\mathfrak{R}^k} |\widehat{\rho}_t^i(s) - \rho(s)| \, ds = O(t^{-\delta}) \quad \text{as} \quad t \to \infty, \tag{3.15}$$

for some $\delta > 0$. For instance, under suitable conditions on the density ρ, the kernel estimate satisfies (3.15) with $\delta = 2/5$ (see Theorem D.4 in Appendix D). Another example is given in Appendix D.2.3 for a particular density belonging to a parametric family of densities.

Now, for $p > 1$ given in Assumption 3.4, let $q > 1$ be such that $1/p + 1/q = 1$. Then, from Remark D.2 (Appendix D) we have

$$E\|\widehat{\rho}_t^i - \rho\|_{L_1}^q = E\left(\int_{\mathfrak{R}^k} |\widehat{\rho}_t^i(s) - \rho(s)| \, ds \right)^q = O(t^{-\delta}) \quad \text{as} \quad t \to \infty. \tag{3.16}$$

Under Assumption 3.4(b), the class of densities $\mathscr{D}$ is a closed and convex subset of L_q (see Proposition 2.1). Moreover (by (2.36)) we have that the projection of each density $\widehat{\rho}_t^i(\cdot), t \in \mathbb{N}$, on $\mathscr{D}$ is well defined, that is, there exists a unique density $\rho_t^i(\cdot) \in \mathscr{D}$ for each $i \in \{1, 2\}$ such that

$$\|\rho_t^i - \widehat{\rho}_t^i\|_{L_1} = \inf_{\sigma \in \mathscr{D}} \|\sigma - \widehat{\rho}_t^i\|_{L_1}.$$

Hence, the density estimators ρ_t^i, $i \in \{1, 2\}$, are defined by

$$\rho_t^i := \operatorname*{argmin}_{\sigma \in \mathscr{D}} \|\sigma - \widehat{\rho}_t^i\|_{L_1}. \tag{3.17}$$

The next result establishes the convergence of the densities $\{\rho_t^i(\cdot)\}$ to $\rho(\cdot)$ for $i \in \{1, 2\}$; the proof is obtained by applying similar arguments to those of Lemma 2.1.

Lemma 3.1. *Suppose Assumption 3.4 holds. Then*

$$E\|\rho_t^i - \rho\|^q = O(t^{-\delta}) \quad \text{as} \quad t \to \infty,$$

where $\delta > 0$ is as in (3.16) and

$$\|\sigma\| := \sup_{(x,a,b) \in \mathbb{K}} \frac{1}{W(x)} \int_{\mathfrak{R}^k} W[F(x,a,b,s)] |\sigma(s)| \, ds, \tag{3.18}$$

for any measurable function $\sigma : \mathfrak{R}^k \to \mathfrak{R}$.

Note that Condition **D.2** guarantees $\|\sigma\| < \infty$ for any density σ in $\mathscr{D}$.

3.3.1 Average Optimal Pair of Strategies

For a nondecreasing sequence of discount factors $\{\alpha_t\}$, we denote by $\kappa(n), n \in \mathbb{N}$, the times this sequence changes its values among the first n terms; that is,

$$\kappa(n) = |\{\alpha_t : t = 1, \ldots, n\}| - 1,$$

where $|C|$ denotes the cardinality of the set C.

We choose a sequence $\{\hat{\alpha}_t\}$ with the following properties:

F.1 $(1 - \hat{\alpha}_t)^{-1} = O(t^v)$ as $t \to \infty$, with $v \in (0, \delta/3q)$.

F.2 $\displaystyle\lim_{n \to \infty} \frac{\kappa(n)}{n} = 0.$

Observe that Condition F.2 implies that, for each $\varepsilon > 0$, $\kappa(n)$ could only be greater than εn for finitely many values of n. Thus, $\{\hat{\alpha}_t\}$ remains constant for long time periods.

An example of a sequence $\{\hat{\alpha}_t\}$ satisfying Conditions F.1 and F.2 is the following.

Example 3.1. Let δ and q be given real numbers satisfying (3.16), and $\{\alpha_t\}$ be the sequence defined as

$$\alpha_t := 1 - \frac{1}{t^v},$$

for some $v \in (0, \delta/3q)$. We define the sequence $\{\hat{\alpha}_t\}$ by

$$\hat{\alpha}_t = \alpha_k \quad \text{if} \quad \frac{(k-1)k}{2} \leq t < \frac{k(k+1)}{2}, \quad k = 2, 3, \ldots$$

Then, since $k \leq t$, straightforward calculations shows that

$$(1 - \hat{\alpha}_t)^{-1} = (1 - \alpha_k)^{-1} = k^v = O(t^v).$$

Moreover, for $n \in \mathbb{N}$,

$$\kappa(n) = (k-2) \quad \text{if} \quad \frac{(k-1)k}{2} \leq n < \frac{k(k+1)}{2}.$$

Therefore, $\kappa(n) < \sqrt{2n}$ and

$$\frac{\kappa(n)}{n} \to 0.$$

Hence, Condition F.2 holds. ∎

From Assumption 3.1, because $\rho_t^i \in \mathscr{D}, t \in \mathbb{N}_0$, there exist functions $V_{\hat{\alpha}_t}^{\rho_t^i}, i = 1, 2,$ and sequences of stochastic kernels $\{\varphi_t^1\} \subset \Phi^1, \{\varphi_t^2\} \subset \Phi^2$, such that, for all $x \in X$, $\varphi_t^1(x) \in \mathbb{A}(x)$ and $\varphi_t^2(x) \in \mathbb{B}(x)$ satisfy (see Remark 3.2 and (3.12))

$$V_{\hat{\alpha}_t}^{\rho_t^1}(x) = T_{(\hat{\alpha}_t, \rho_t^1)} V_{\hat{\alpha}_t}^{\rho_t^1}(x)$$

$$= \min_{\varphi^2 \in \mathbb{B}(x)} \left[r(x, \varphi_t^1, \varphi^2) + \hat{\alpha}_t \int_{\Re^k} V_{\hat{\alpha}_t}^{\rho_t^1}(F(x, \varphi_t^1, \varphi^2, s)) \rho_t^1(s) ds \right] \quad (3.19)$$

and

$$V_{\hat{\alpha}_t}^{\rho_t^2}(x) = T_{(\hat{\alpha}_t, \rho_t^2)} V_{\hat{\alpha}_t}^{\rho_t^2}(x)$$

$$= \max_{\varphi^1 \in \mathbb{A}(x)} \left[r(x, \varphi^1, \varphi_t^2) + \hat{\alpha}_t \int_{\Re^k} V_{\hat{\alpha}_t}^{\rho_t^2}(F(x, \varphi^1, \varphi_t^2, s)) \rho_t^2(s) ds \right]. \quad (3.20)$$

We state our main result as follows.

Theorem 3.5. *Suppose that Assumptions 3.1, 3.2, and 3.4 hold. Then the strategies* $\pi_*^1 = \{\varphi_t^1\} \in \Pi^1$ *and* $\pi_*^2 = \{\varphi_t^2\} \in \Pi^2$ *are average optimal for player 1 and 2, respectively, that is,*

$$j^* = \inf_{\pi^2 \in \Pi^2} J(x, \pi_*^1, \pi^2) = \sup_{\pi^1 \in \Pi^1} J(x, \pi^1, \pi_*^2) \quad \forall x \in X.$$

3.3.2 Proof of Theorem 3.5

For the proof of our result we need some notation and several simple but useful facts. From (3.19) and (3.20), for each $t \in \mathbb{N}$,

$$j_{\hat{\alpha}_t}^{\rho_t^2} + h_{\hat{\alpha}_t}^{\rho_t^2}(x) = \max_{\varphi^1 \in \mathbb{A}(x)} \left[r(x, \varphi^1, \varphi_t^2) + \hat{\alpha}_t \int_{\Re^k} h_{\hat{\alpha}_t}^{\rho_t^2}(F(x, \varphi^1, \varphi_t^2, s)) \rho_t^2(s) ds \right] \quad (3.21)$$

and

$$j_{\hat{\alpha}_t}^{\rho_t^1} + h_{\hat{\alpha}_t}^{\rho_t^1}(x) = \min_{\varphi^2 \in \mathbb{B}(x)} \left[r(x, \varphi_t^1, \varphi^2) + \hat{\alpha}_t \int_{\Re^k} h_{\hat{\alpha}_t}^{\rho_t^1}(F(x, \varphi_t^1, \varphi^2, s)) \rho_t^1(s) ds \right], \quad (3.22)$$

for all $x \in X$. Thus, from (3.7) in Theorem 3.3 (see (3.9))

$$j^* = \lim_{t \to \infty} j_{\hat{\alpha}_t}^{\rho}. \quad (3.23)$$

Again, from inequalities (3.10) and (3.14) (see (2.24) and (2.21) for the discounted case), the relations (2.26) and (2.27) hold:

$$\sup_{t\in\mathbb{N}_0} E_x^{\pi^1,\pi^2}[W^p(x_t)] < \infty \quad \text{and} \quad \sup_{t\in\mathbb{N}_0} E_x^{\pi^1,\pi^2}[W(x_t)] < \infty \qquad (3.24)$$

for each pair $(\pi^1,\pi^2) \in \Pi^1 \times \Pi^2$ and $x \in X$. Now, the geometric ergodicity (3.2) implies that

$$|h_\alpha(x)| \le \frac{R}{1-\gamma}[1+W(z)]W(x) \ \forall x \in X, \alpha \in (0,1).$$

Hence,

$$\sup_{\alpha\in(0,1)} \|h_\alpha\|_W < \infty. \qquad (3.25)$$

See, for instance, [31, Lemma 10.4.2, p. 138].

Remark 3.3 (See Remark 2.3). Let $\hat{\alpha}_t$ be the sequence satisfying the Conditions F.1 and F.2, and d be the constant in (3.10). For each $t \in \mathbb{N}_0$, define $\gamma_t := (1+\hat{\alpha}_t)/2 \in (\hat{\alpha}_t, 1)$, $e_t := d(\gamma_t/\hat{\alpha}_t - 1)^{-1}$, and the function $W_t(x) := W(x) + e_t$ for $x \in X$. Now, consider the space $\mathbb{B}_{W_t}(X)$ of functions $u : X \to \mathfrak{R}$ with finite W_t-norm, that is

$$\|u\|_{W_t} := \sup_{x\in X} \frac{|u(x)|}{W_t(x)} < \infty \quad \text{for all } t \in \mathbb{N}_0,$$

and observe that for each $t \in \mathbb{N}_0$ this norm is equivalent to the W-norm since

$$\|u\|_{W_t} \le \|u\|_W \le l_t \|u\|_{W_t}, \qquad (3.26)$$

where

$$l_t := 1 + \frac{2\hat{\alpha}_t d}{1-\hat{\alpha}_t}. \qquad (3.27)$$

Then, for any density $\sigma \in \mathscr{D}$, (3.10) implies that the function W_t satisfies the inequality

$$\hat{\alpha}_t \int_{\mathfrak{R}^k} W_t[F(x,a,b,s)]\sigma(s)ds \le \gamma_t W_t(x), \quad \forall(x,a,b) \in \mathbb{K}, t \in \mathbb{N}_0.$$

Hence, from (2.12), $T_{(\hat{\alpha}_t,\sigma)}$ is a contraction with respect to the W_t-norm with modulus γ_t, i.e.,

$$\left\|T_{(\hat{\alpha}_t,\sigma)}v - T_{(\hat{\alpha}_t,\sigma)}u\right\|_{W_t} \le \gamma_t \|v-u\|_{W_t} \quad \forall v,u \in \mathbb{B}_W, t \in \mathbb{N}_0. \qquad (3.28)$$

Lemma 3.2. *Suppose that Assumptions 3.1, 3.2, and 3.4 hold. Then, for each $x \in X$, $(\pi^1,\pi^2) \in \Pi^1 \times \Pi^2$, and $i \in \{1,2\}$, as $t \to \infty$*
  (a) $E_x^{\pi^1,\pi^2}\|h_{\hat{\alpha}_t}^\rho - h_{\hat{\alpha}_t}^{\rho^i}\|_W^q \to 0$,
  (b) $E_x^{\pi^1,\pi^2}\left[\|h_{\hat{\alpha}_t}^\rho - h_{\hat{\alpha}_t}^{\rho^i}\|_W W(x_t)\right] \to 0$.

Proof. (a) Observe that for each $t \in \mathbb{N}_0$ and $i \in \{1,2\}$, $\|h_{\hat{\alpha}_t} - h_{\hat{\alpha}_t}^{\rho^i}\|_W \le 2\|V_{\hat{\alpha}_t}^\rho - V_{\hat{\alpha}_t}^{\rho^i}\|_W$. Hence, from (3.26), it is enough to prove that

$$l_t^q E_x^{\pi^1,\pi^2} ||V_{\hat{\alpha}_t}^\rho - V_{\hat{\alpha}_t}^{\rho_t^i}||_{W_t}^q \to 0, \text{ as } t \to \infty. \tag{3.29}$$

From (3.12), (3.19), and (3.20), for each $t \in \mathbb{N}_0$, $T_{(\hat{\alpha}_t,\rho)}V_{\hat{\alpha}_t}^\rho = V_{\hat{\alpha}_t}^\rho$ and $T_{(\hat{\alpha}_t,\rho_t^i)}V_{\hat{\alpha}_t}^{\rho_t^i} = V_{\hat{\alpha}_t}^{\rho_t^i}$. Thus, from (3.28), for all $t \in \mathbb{N}_0$ we have

$$||V_{\hat{\alpha}_t}^\rho - V_{\hat{\alpha}_t}^{\rho_t^i}||_{W_t} \le ||T_{(\hat{\alpha}_t,\rho)}V_{\hat{\alpha}_t}^\rho - T_{(\hat{\alpha}_t,\rho_t^i)}V_{\hat{\alpha}_t}^\rho||_{W_t} + \gamma_t||V_{\hat{\alpha}_t}^\rho - V_{\hat{\alpha}_t}^{\rho_t^i}||_{W_t},$$

which in turn implies

$$l_t||V_{\hat{\alpha}_t}^\rho - V_{\hat{\alpha}_t}^{\rho_t^i}||_{W_t} \le \frac{l_t}{1-\gamma_t}||T_{(\hat{\alpha}_t,\rho)}V_{\hat{\alpha}_t}^\rho - T_{(\hat{\alpha}_t,\rho_t^i)}V_{\hat{\alpha}_t}^\rho||_{W_t}. \tag{3.30}$$

Now, from (3.3), (3.18), and using the fact $[W_t(\cdot)]^{-1} < [W(\cdot)]^{-1}$, we obtain

$$||T_{(\hat{\alpha}_t,\rho)}V_{\hat{\alpha}_t}^\rho - T_{(\hat{\alpha}_t,\rho_t^i)}V_{\hat{\alpha}_t}^\rho||_{W_t}$$

$$\le \hat{\alpha}_t \sup_{(x,a,b)\in\mathbb{K}} W_t^{-1}(x)\int_{\Re^k} V_{\hat{\alpha}_t}^\rho(F(x,a,b,s))|\rho(s)-\rho_t^i(s)|ds$$

$$\le \frac{M\alpha_t}{1-\hat{\alpha}_t} \sup_{(x,a,b)\in\mathbb{K}} W^{-1}(x)\int_{\Re^k} W(F(x,a,b,s))|\rho(s)-\rho_t^i(s)|ds$$

$$\le \frac{M}{1-\hat{\alpha}_t}||\rho-\rho_t^i||. \tag{3.31}$$

On the other hand, note that $\hat{\alpha}_t$ and γ_t satisfy the relation

$$\frac{1}{(1-\gamma_t)(1-\hat{\alpha}_t)^2} = O(t^{3\nu}) \text{ as } t \to \infty. \tag{3.32}$$

Combining (3.30)–(3.32), and using (3.27) we get

$$l_t^q||V_{\hat{\alpha}_t}^\rho - V_{\hat{\alpha}_t}^{\rho_t^i}||_{W_t}^q \le M^q\left[\frac{1}{(1-\gamma_t)(1-\hat{\alpha}_t)} + \frac{2d}{(1-\gamma_t)(1-\hat{\alpha}_t)^2}\right]^q ||\rho-\rho_t^i||^q$$

$$= M^q O(t^{3q\nu})||\rho-\rho_t^i||^q \text{ as } t \to \infty. \tag{3.33}$$

Finally, (3.29) follows from Lemma 3.1 taking expectation with respect to $P_x^{\pi^1,\pi^2}$ on both sides of (3.33), after noting that $3\nu p' < \delta$ (see Condition F.1).

(b) To prove the statement (b), let $\bar{C} := \left(E_x^{\pi^1,\pi^2}[W^p(x_t)]\right)^{1/p} < \infty$ (see (3.24)). Then, Holder's inequality and part (a) of this lemma yield

$$E_x^{\pi^1,\pi^2}\|h_{\hat{\alpha}_t} - h_{\hat{\alpha}_t}^{p_i}\|_W W(x_t) \leq \check{C}\left(E_x^{\pi^1,\pi^2}\left[\|h_{\hat{\alpha}_t}^{p} - h_{\hat{\alpha}_t}^{p_i}\|_W^q\right]\right)^{1/q} \to 0, \text{ as } t \to \infty.$$

This completes the proof of Lemma 3.2. ∎

Proof (Proof of Theorem 3.5). To prove the theorem it is enough to show that

$$J(\pi_*^1,\pi^2,x) \geq j^* \quad \forall \pi^2 \in \Pi^2, x \in X, \tag{3.34}$$

and

$$J(\pi^1,\pi_*^2,x) \leq j^* \quad \forall \pi^1 \in \Pi^1, x \in X, \tag{3.35}$$

which imply the optimality of $\pi_*^1 = \{\varphi_t^1\}$ and $\pi_*^2 = \{\varphi_t^2\}$, respectively. We first prove the optimality of π_*^2.

Let $\pi^1 = \{\pi_t^1\} \in \Pi^1$ be an arbitrary strategy for player 1. Define

$$\eta_t := r(x_t,\pi_t^1,\varphi_t^2) + \hat{\alpha}_t\int_{\mathfrak{R}^k} h_{\hat{\alpha}_t}^{p}[F(x_t,\pi_t^1,\varphi_t^2,s)]\rho(s)ds - j_{\hat{\alpha}_t}^{p} - h_{\hat{\alpha}_t}^{p}(x_t)$$

$$= r(x_t,\pi_t^1,\varphi_t^2) + \hat{\alpha}_t E_x^{\pi^1,\pi_*^2}\left[h_{\hat{\alpha}_t}^{p}(x_{t+1}) \mid h_t\right] - j_{\hat{\alpha}_t}^{p} - h_{\hat{\alpha}_t}^{p}(x_t). \tag{3.36}$$

Thus

$$E_x^{\pi^1,\pi_*^2} r(x_t,a_t,b_t) = j_{\hat{\alpha}_t}^{p} + E_x^{\pi^1,\pi_*^2}\left[h_{\hat{\alpha}_t}^{p}(x_t) - \hat{\alpha}_t h_{\hat{\alpha}_t}^{p}(x_{t+1})\right] + E_x^{\pi^1,\pi_*^2}\eta_t.$$

We now proceed to prove that

$$J(\pi^1,\pi_*^2,x) = \liminf_{n\to\infty}\frac{1}{n}E_x^{\pi^1,\pi_*^2}\sum_{t=0}^{n-1} r(x_t,a_t,b_t) \leq j^* \quad \forall x \in X.$$

To do this, write

$$n^{-1}E_x^{\pi^1,\pi_*^2}\left[\sum_{t=0}^{n-1} r(x_t,a_t,b_t)\right] = I_1(n) + I_2(n) + I_3(n) + I_4(n), \tag{3.37}$$

where, for $n \geq k \geq 1$,

$$I_1(n) := n^{-1}\sum_{t=0}^{n-1} j_{\hat{\alpha}_t}^{p};$$

$$I_2(n) := n^{-1}E_x^{\pi^1,\pi_*^2}\left[\sum_{t=0}^{k-1}\left(h_{\hat{\alpha}_t}^{p}(x_t) - \hat{\alpha}_t h_{\hat{\alpha}_t}^{p}(x_{t+1})\right)\right]$$

$$I_3(n) := n^{-1} E_x^{\pi^1, \pi_*^2} \left[\sum_{t=k}^{n-1} \left(h_{\hat{\alpha}_t}^{\rho}(x_t) - \hat{\alpha}_t h_{\hat{\alpha}_t}^{\rho}(x_{t+1}) \right) \right]$$

$$I_4(n) := n^{-1} E_x^{\pi^1, \pi_*^2} \left[\sum_{t=0}^{n-1} \eta_t \right].$$

From (3.23) we have

$$I_1(n) \to j^* \text{ as } n \to \infty. \tag{3.38}$$

We also have

$$I_2(n) \to 0, \text{ as } n \to \infty. \tag{3.39}$$

On the other hand, let $C' < \infty$ be a constant such that $E_x^{\pi^1, \pi_*^2}[h_\alpha(x_t)] < C'$ for all $\alpha \in (0,1)$ and $t \in \mathbb{N}_0$ (see (3.24) and (3.25)). Recall that the sequence $\{\hat{\alpha}_t\}$ satisfies Conditions F.1 and F.2. Thus, let $\alpha_1^*, \alpha_2^*, \ldots, \alpha_{\kappa(n)}^*$ be the different values in the set $\{\hat{\alpha}_t : 1 \le t \le n\}$ of $\hat{\alpha}_t$. Then, using that $\{\hat{\alpha}_t\}$ is a nondecreasing sequence, for any $n \ge k \ge 1$, we have

$$0 \le |I_3(n)| \le n^{-1} \left| E_x^{\pi^1, \pi_*^2} \left[\sum_{t=k}^{n-1} \left(h_{\hat{\alpha}_t}^{\rho}(x_t) - \hat{\alpha}_t h_{\hat{\alpha}_t}^{\rho}(x_t) \right) \right] \right|$$

$$+ n^{-1} \left| E_x^{\pi^1, \pi_*^2} \left[\sum_{t=k}^{n-1} \hat{\alpha}_t \left(h_{\hat{\alpha}_t}^{\rho}(x_t) - h_{\hat{\alpha}_t}^{\rho}(x_{t+1}) \right) \right] \right|$$

$$\le (1 - \alpha_k)C' + n^{-1}C' \sum_{i=1}^{\kappa(n)} \alpha_i^* \le (1 - \alpha_k)C' + C'\kappa(n)n^{-1} \tag{3.40}$$

Hence, since k is arbitrary and $\hat{\alpha}_t \uparrow 1$ as $t \to \infty$, the Condition F.2 implies that

$$\lim_{n \to \infty} I_3(n) = 0. \tag{3.41}$$

Now we will prove that

$$\lim_{n \to \infty} I_4(n) \le 0. \tag{3.42}$$

Adding and subtracting in (3.36) the terms

$$\hat{\alpha}_t \int_{\Re^k} h_{\hat{\alpha}_t}^{\rho_t^2} [F(x_t, \pi_t^1, \varphi_t^2, s)] \rho(s) ds \text{ and } \hat{\alpha}_t \int_{\Re^k} h_{\hat{\alpha}_t}^{\rho_t^2} [F(x_t, \pi_t^1, \varphi_t^2, s)] \rho_t^2(s) ds$$

we get

$$\eta_t = \hat{\alpha}_t \int_{\Re^k} h^{\rho}_{\hat{\alpha}_t}[F(x_t, \pi^1_t, \varphi^2_t, s)]\rho(s)ds - \hat{\alpha}_t \int_{\Re^k} h^{\rho^2_t}_{\hat{\alpha}_t}[F(x_t, \pi^1_t, \varphi^2_t, s)]\rho(s)ds$$

$$+ \hat{\alpha}_t \int_{\Re^k} h^{\rho^2_t}_{\hat{\alpha}_t}[F(x_t, \pi^1_t, \varphi^2_t, s)]\rho(s)ds - \hat{\alpha}_t \int_{\Re^k} h^{\rho^2_t}_{\hat{\alpha}_t}[F(x_t, \pi^1_t, \varphi^2_t, s)]\rho^2_t(s)ds$$

$$+ r(x_t, \pi^1_t, \varphi^2_t) + \hat{\alpha}_t \int_{\Re^k} h^{\rho^2_t}_{\hat{\alpha}_t}[F(x_t, \pi^1_t, \varphi^2_t, s)]\rho^2_t(s)ds - j^{\rho}_{\hat{\alpha}_t} - h^{\rho}_{\hat{\alpha}_t}(x_t). \qquad (3.43)$$

On the other hand, since $h^{\rho}_{\hat{\alpha}_t}, h^{\rho^2_t}_{\hat{\alpha}_t} \in \mathbb{B}_W$, from (3.10) and using the fact that $|u(\cdot)| \le \|u\|_W W(\cdot)$ for any $u \in \mathbb{B}_W$, we have

$$\hat{\alpha}_t \int_{\Re^k} h^{\rho}_{\hat{\alpha}_t}[F(x_t, \pi^1_t, \varphi^2_t, s)]\rho(s)ds - \hat{\alpha}_t \int_{\Re^k} h^{\rho^2_t}_{\hat{\alpha}_t}[F(x_t, \pi^1_t, \varphi^2_t, s)]\rho(s)ds$$

$$\le \hat{\alpha}_t \|h^{\rho}_{\hat{\alpha}_t} - h^{\rho^2_t}_{\hat{\alpha}_t}\|_W [\beta W(x_t) + d]. \qquad (3.44)$$

Similarly, from (3.18),

$$\hat{\alpha}_t \int_{\Re^k} h^{\rho^2_t}_{\hat{\alpha}_t}[F(x_t, \pi^1_t, \varphi^2_t, s)]\rho(s)ds - \hat{\alpha}_t \int_{\Re^k} h^{\rho^2_t}_{\hat{\alpha}_t}[F(x_t, \pi^1_t, \varphi^2_t, s)]\rho^2_t(s)ds$$

$$\le \hat{\alpha}_t W(x_t) \|h^{\rho^2_t}_{\hat{\alpha}_t}\|_W [W(x_t)]^{-1} \int_{\Re^k} W[F(x_t, \pi^1_t, \varphi^2_t, s)] |\rho(s) - \rho^2_t(s)| ds$$

$$\le \hat{\alpha}_t W(x_t) \|h^{\rho^2_t}_{\hat{\alpha}_t}\|_W \|\rho - \rho^2_t\| \qquad (3.45)$$

Then, combining (3.43), (3.45), and (3.21), we get

$$\eta_t \le \hat{\alpha}_t \|h^{\rho}_{\hat{\alpha}_t} - h^{\rho^2_t}_{\hat{\alpha}_t}\|_W [\beta W(x_t) + d] + \hat{\alpha}_t W(x_t) \|h^{\rho^2_t}_{\hat{\alpha}_t}\|_W \|\rho - \rho^2_t\|$$

$$+ \max_{\varphi^1 \in \mathbb{A}(x)} \left\{ r(x_t, \varphi^1, \varphi^2_t) + \hat{\alpha}_t \int_{\Re^k} h^{\rho^2_t}_{\hat{\alpha}_t}[F(x_t, \varphi^1, \varphi^2_t, s)]\rho^2_t(s)ds \right\} - j^{\rho}_{\hat{\alpha}_t} - h^{\rho}_{\hat{\alpha}_t}(x_t)$$

$$= \hat{\alpha}_t \|h^{\rho}_{\hat{\alpha}_t} - h^{\rho^2_t}_{\hat{\alpha}_t}\|_W [\beta W(x_t) + d] + \hat{\alpha}_t W(x_t) \|h^{\rho^2_t}_{\hat{\alpha}_t}\|_W \|\rho - \rho^2_t\|$$

$$+ j^{\rho^2_t}_{\hat{\alpha}_t} + h^{\rho^2_t}_{\hat{\alpha}_t}(x_t) - j^{\rho}_{\hat{\alpha}_t} - h^{\rho}_{\hat{\alpha}_t}(x_t). \qquad (3.46)$$

Additionally, Lemma 3.2 implies

$$\lim_{t \to \infty} E_x^{\pi^1, \pi_*^2} \left[\hat{\alpha}_t \| h_{\hat{\alpha}_t}^{\rho} - h_{\hat{\alpha}_t}^{\rho_t^2} \|_W (\beta W(x_t) + d) \right] = 0, \tag{3.47}$$

whereas (3.3) together with Condition F.1 yield

$$\| h_{\hat{\alpha}_t}^{\rho_t^2} \|_W \le 2 \| V_{\hat{\alpha}_t}^{\rho_t^2} \|_W \le \frac{2M}{1 - \hat{\alpha}_t} = O(t^v) \text{ as } t \to \infty.$$

Therefore, since $v < \delta/q$, taking expectation and applying Hölder's inequality we obtain

$$0 \le E_x^{\pi^1, \pi_*^2} \left[W(x_t) \| h_{\hat{\alpha}_t}^{\rho_t^2} \|_W \| \rho - \rho_t^2 \| \right]$$

$$\le \hat{C} \left([O(t^v)]^q E_x^{\pi^1, \pi_*^2} \| \rho - \rho_t^2 \|^q \right)^{1/q}$$

$$= \left[O(t^{vq - \gamma}) \right]^{1/q} \to 0 \text{ as } t \to \infty, \tag{3.48}$$

where $\hat{C} := \left(E_x^{\pi^1, \pi_*^2} W^p(x_t) \right)^{1/p} < \infty$ (see (3.24)).

On the other hand, observe that

$$\left| j_{\hat{\alpha}_t}^{\rho_t^2} - j_{\hat{\alpha}_t}^{\rho} \right| \le (1 - \hat{\alpha}_t) \| V_{\hat{\alpha}_t}^{\rho_t^2} - V_{\hat{\alpha}_t}^{\rho} \|_W W(z),$$

and

$$\left| h_{\hat{\alpha}_t}^{\rho_t^2}(x_t) - h_{\hat{\alpha}_t}^{\rho}(x_t) \right| \le \| h_{\hat{\alpha}_t}^{\rho_t^2} - h_{\hat{\alpha}_t}^{\rho} \|_W W(x_t).$$

Hence, (3.33) and Lemma 3.2 yield

$$\lim_{t \to \infty} E_x^{\pi^1, \pi_*^2} \left(j_{\hat{\alpha}_t}^{\rho_t^2} - j_{\hat{\alpha}_t}^{\rho} \right) = 0 \tag{3.49}$$

and

$$\lim_{t \to \infty} E_x^{\pi^1, \pi_*^2} \left(h_{\hat{\alpha}_t}^{\rho_t^2}(x_t) - h_{\hat{\alpha}_t}^{\rho}(x_t) \right) = 0. \tag{3.50}$$

Therefore, combining (3.46)–(3.50) we obtain (3.42). Finally, letting $n \to \infty$ in (3.37), from (3.1) and (3.38)–(3.42), we get

$$J(\pi^1, \pi_*^2, x) \le j^*, \quad \forall \pi^1 \in \Pi^1, x \in X,$$

that is, (3.35) holds.

The proof of (3.34) follows by applying similar arguments as the proof of (3.35) given the symmetric role that the players have in the game. Indeed, let $\pi^2 = \{\pi_t^2\} \in$

Π^2 be an arbitrary strategy for player 2. Defining accordingly η_t, $I_1(n)$, $I_2(n)$, $I_3(n)$, and $I_4(n)$ with $E_x^{\pi_*^1,\pi^2}$ and ρ_t^1 instead of $E_x^{\pi^1,\pi_*^2}$ and ρ_t^2, respectively, it is easy to see that, as $n \to \infty$,

$$I_1(n) \to j^*, \quad I_2(n) \to 0, \quad \text{and} \quad I_3(n) \to 0.$$

Observe that in this case instead of (3.42) we now obtain

$$\lim_{n\to\infty} I_4(n) \geq 0.$$

In fact, this follows from (3.47) to (3.50) and noting that (as in (3.43)–(3.45))

$$\eta_t \geq -\left\{ \hat{\alpha}_t \|h_{\hat{\alpha}_t}^{\rho} - h_{\hat{\alpha}_t}^{\rho_t^1}\|_W [\beta W(x_t) + d] + \hat{\alpha}_t W(x_t)\|h_{\hat{\alpha}_t}^{\rho_t^1}\|_W \|\rho - \rho_t^1\| \right\}$$

$$+ \min_{\varphi^2 \in \mathbb{B}(x)} \left[r(x_t, \mu_t^*, \varphi^2) + \hat{\alpha}_t \int_{\Re^k} h_{\hat{\alpha}_t}^{\rho_t^1} (F(x_t, \mu_t^*, \varphi^2, s)) \rho_t^1(s) ds \right]$$

$$- j_{\hat{\alpha}_t}^{\rho} - h_{\hat{\alpha}_t}^{\rho} (x_t)$$

$$\geq -\left\{ \hat{\alpha}_t \|h_{\hat{\alpha}_t}^{\rho} - h_{\hat{\alpha}_t}^{\rho_t^1}\|_W [\beta W(x_t) + d] + \hat{\alpha}_t W(x_t)\|h_{\hat{\alpha}_t}^{\rho_t^1}\|_W \|\rho - \rho_t^1\| \right\}$$

$$+ j_{\hat{\alpha}_t}^{\rho_t^1} + h_{\hat{\alpha}_t}^{\rho_t^1} (x_t) - j_{\hat{\alpha}_t}^{\rho} - h_{\hat{\alpha}_t}^{\rho} (x_t),$$

where the last inequality follows from (3.22).

In conclusion, we obtain

$$J(\pi_*^1, \pi^2, x) \geq j^* \quad \forall \pi^2 \in \Pi^2, x \in X,$$

which proves the desired result. ∎

Remark 3.4 (Bounded Payoffs). The average optimality of the strategies $\pi_*^1 = \{\varphi_t^1\} \in \Pi^1$ and $\pi_*^2 = \{\varphi_t^2\} \in \Pi^2$ in Theorem 3.5 is based essentially on the Conditions F.1 and F.2 of the sequence of discount factors $\{\hat{\alpha}_t\}$, and their link with the rate of convergence of the density estimator (see Lemma 3.1 and Condition F.1). As the discounted criterion (see Remark 2.8), if the payoff function r is bounded we can take $W \equiv 1$ and any L_1-consistent density estimator with a suitable rate of convergence can be used to construct the strategies. That is, once fixed $\delta > 0$ such that, for $i = 1, 2$,

$$E \|\rho_t^i - \rho\|_{L_1} = O(t^{-\delta}) \text{ as } t \to \infty,$$

we define the sequence $\{\hat{\alpha}_t\}$ of discount factors satisfying the Conditions F.1 and F.2 (see, e.g., Example 3.1). In fact, it is possible to prove, in specific examples, that Theorem 3.5 remains to be valid for sequences of discount factors that converge enough slowly to one. Hence, Conditions F.1 and F.2 can be weakened.

Furthermore, the ergodicity property given in (3.2), which is a consequence of Assumption 3.2, becomes in the well-known geometric ergodicity in the total variation norm (see, e.g., [19, 28]).

Chapter 4
Empirical Approximation-Estimation Algorithms in Markov Games

This chapter proposes an empirical approximation-estimation algorithm in difference equation game models (see Sect. 1.1.1) whose evolution is given by

$$x_{t+1} = F(x_t, a_t, b_t, \xi_t), \quad t \in \mathbb{N}_0, \tag{4.1}$$

where $\{\xi_t\}$ is a sequence of observable i.i.d. random variables defined on a probability space $(\Omega, \mathscr{F}, P)$, taking values in an arbitrary Borel space S, with common unknown distribution $\theta \in \mathbb{P}(S)$.

Note that unlike the previous chapters, we are now considering an arbitrary distribution θ, which not necessarily has a density. Our objective in this scenario is to use an empirical procedure to estimate θ that in turn defines an algorithm to approximate the value of the game as well as an optimal pair of strategies. This is done for both, discounted and average criteria, by applying the following approach.

As was seen previously, the study of Markov games with discounted payoffs is analyzed by means of the Shapley equation $T_{(\alpha,\theta)} V_\alpha^\theta = V_\alpha^\theta$, where V_α^θ is the value of the game and $T_{(\alpha,\theta)}$ is a minimax (maximin) operator. Then, in the setting of θ completely known, under suitable condition, a stationary optimal pair of strategies $(\pi_*^1, \pi_*^2) = (\{\varphi_*^1\}, \{\varphi_*^2\}) \in \Pi_s^1 \times \Pi_s^2$ can be computed. Now, assuming unknown θ, given a sample $\bar{\xi}_n = (\xi_0, \xi_1, \dots, \xi_{n-1})$, the corresponding empirical measure $\theta_n(\cdot) = \theta_n(\cdot; \bar{\xi}_n)$ defines a random operator $T_{(\alpha,\theta_n)}$ and an empirical value $V_\alpha^{\theta_n}$ satisfying $T_{(\alpha,\theta_n)} V_\alpha^{\theta_n} = V_\alpha^{\theta_n}$. So, for each $n \in \mathbb{N}$, it is possible to get an optimal pair $(\{\varphi_n^1\}, \{\varphi_n^2\}) \in \Pi_s^1 \times \Pi_s^2$ for the θ_n-empirical game. Then, under suitable conditions, we prove the convergence $V_\alpha^{\theta_n} \to V_\alpha^\theta$ and the existence of a limit point $(\varphi_\infty^1, \varphi_\infty^2)$ of $\{(\varphi_n^1, \varphi_n^2)\}$, which defines a stationary optimal pair of strategies $(\pi_\infty^1, \pi_\infty^2) = (\{\varphi_\infty^1\}, \{\varphi_\infty^2\})$ for the original game. It is worth observing that by the

J. A. Minjárez-Sosa, *Zero-Sum Discrete-Time Markov Games with Unknown Disturbance Distribution*, SpringerBriefs in Probability and Mathematical Statistics, https://doi.org/10.1007/978-3-030-35720-7_4

randomness of the operator $T_{(\alpha,\theta_n)}$ as well as of the functions $V_\alpha^{\theta_n}$, the pair $\left(\varphi_\infty^1, \varphi_\infty^2\right)$ is a random variable. Hence, as part of our approach, we have to prove that its expectation determines an optimal (nonrandom) pair of strategies.

Once the discounted case is analyzed, the average payoff criterion is studied by means of a combination of the VDFA and the empirical estimation procedure.

Our approach, in addition to providing a more general method for estimating value functions and for the construction of optimal strategies, can be seen as an approximation method in cases where θ can be known but difficult to handle, i.e., θ is replaced by a simpler distribution, namely, the empirical distribution θ_n.

Throughout the chapter, a.s. means "almost surely" with respect to the underlying probability measure P.

4.1 Assumptions and Preliminary Results

The theory on the empirical procedures for discounted and average payoff games will be developed in the setting imposed by the Assumption 3.1 (see Assumptions 2.1, 2.4, and 2.8(b), and Remark 2.4(a)), together with the ergodicity condition in Assumption 3.2. For ease reference, we rewrite them in terms of the distribution θ and the difference equation game model.

Assumption 4.1 *(a) The multifunctions $x \longmapsto A(x)$ and $x \longmapsto B(x)$ are compact-valued and continuous.*

(b) The payoff function r is continuous on $\mathbb{K}$, and there exist a continuous function $W : X \to [1, \infty)$ and a constant $M > 0$ such that $0 \le r(x,a,b) \le MW(x)$ for all $(x,a,b) \in \mathbb{K}$. Moreover, the function

$$(x,a,b) \longmapsto \int_S W\left[F(x,a,b,s)\right] \theta(ds)$$

is continuous on $\mathbb{K}$.

(c) For each $s \in S$, the function $F(x,a,b,s)$ is continuous in $(x,a,b) \in \mathbb{K}$.

Assumption 4.2 *There exist a measurable function $\lambda : \mathbb{K} \to [0,1]$, a probability measure m^* on X, and a constant $\beta \in (0,1)$ such that:*

(a) $\int_S W[F(x,a,b,s)]\theta(ds) \le \beta W(x) + \lambda(x,a,b)d$ for all $(x,a,b) \in \mathbb{K}$, where

$$d := \int_X W(x)m^*(dx) < \infty;$$

(b) $Q(D|x,a,b) \geq \lambda(x,a,b)m^*(D) \quad \forall D \in \mathscr{B}(X), (x,a,b) \in \mathbb{K}$;

(c) $\int_X \bar{\Lambda}(x)m^*(dx) > 0$, where $\bar{\Lambda}(x) := \inf_{a \in A(x)} \inf_{b \in B(x)} \lambda(x,a,b)$ is assumed to be a measurable function.

Assumption 4.3 *For the constants β and d in Assumption 4.2, the function W satisfies*

$$W[F(x,a,b,s)] \leq \beta W(x) + d \quad \forall (x,a,b,s) \in \mathbb{K} \times S.$$

Remark 4.1. (a) In the particular case of a bounded payoff function r, Assumption 4.3 holds by taking $W \equiv 1$ and $d = 1$ (see Remark 2.2).

(b) Observe that Assumption 4.1(c) implies that the mapping

$$(x,a,b) \longmapsto \int_S v[F(x,a,b,s)]\mu(ds)$$

is continuous on $\mathbb{K}$ for every bounded and continuous function v on X and $\mu \in \mathbb{P}(S)$ (see Remark 2.4(a) or Proposition C.3 in Appendix C).

(c) We consider the following class of probability measures

$$\mathscr{M}(S) := \left\{ \mu \in \mathbb{P}(S) : \int_S W[F(x,a,b,s)]\mu(ds) \leq \beta W(x) + d, \ (x,a,b) \in \mathbb{K} \right\}.$$

Observe that Assumption 4.2(a) implies that $\theta \in \mathscr{M}(S)$, that is

$$\int_S W[F(x,a,b,s)]\theta(ds) \leq \beta W(x) + d$$

for all $(x,a,b) \in \mathbb{K}$. On the other hand, under Assumption 4.3, any probability measure $\mu \in \mathbb{P}(S)$ belongs to $\mathscr{M}(S)$, that is, $\mathscr{M}(S) = \mathbb{P}(S)$.

For each $\mu \in \mathbb{P}(S)$ and $\alpha \in (0,1)$, we define, for $v \in \mathbb{B}_W$ and $x \in X$, the operator (see (2.2) and (2.19))

$$T_{(\alpha,\mu)}v(x) := \inf_{\varphi^2 \in \mathbb{B}(x)} \sup_{\varphi^1 \in \mathbb{A}(x)} \left[r(x,\varphi^1,\varphi^2) + \alpha \int_S v(F(x,\varphi^1,\varphi^2,s))\mu(ds) \right]. \quad (4.2)$$

If Assumption 4.1 holds, standard results in previous chapters (see Theorem 2.6 and Remark 2.4) ensure that, for each $\mu \in \mathscr{M}(S)$, the operator $T_{(\alpha,\mu)}$ maps $\mathbb{C}_W$ into itself, and furthermore the interchange of inf and sup in (4.2) holds:

$$T_{(\alpha,\mu)}v(x) = \sup_{\varphi^1 \in \mathbb{A}(x)} \inf_{\varphi^2 \in \mathbb{B}(x)} \left[r(x,\varphi^1,\varphi^2) + \alpha \int_S v(F(x,\varphi^1,\varphi^2,s))\mu(ds) \right]. \quad (4.3)$$

Moreover, for each $\mu \in \mathbb{P}(S)$ and $\alpha \in (0,1)$, the operator $T_{(\alpha,\mu)}$ has the contraction property given in Remark 2.3. Indeed, let $\bar{W}(x) := W(x) + e$ for $x \in X$, where, for each discount factor $\alpha \in (0,1)$ and arbitrary $\gamma_\alpha \in (\alpha, 1)$,

$$e := d\left(\gamma_\alpha/\alpha - 1\right)^{-1} \tag{4.4}$$

Then (see (2.11)),

$$\alpha \int_S \bar{W}[F(x,a,b,s)]\mu(ds) \le \gamma_\alpha \bar{W}(x) \quad \forall (x,a,b) \in \mathbb{K}, \tag{4.5}$$

and for all $v, u \in \mathbb{B}_W$,

$$\left\| T_{(\alpha,\mu)} v - T_{(\alpha,\mu)} u \right\|_{\bar{W}} \le \gamma_\alpha \|v - u\|_{\bar{W}}. \tag{4.6}$$

Note that the norms $\|\cdot\|_W$ and $\|\cdot\|_{\bar{W}}$ are equivalent since, as in (2.9),

$$\|v\|_{\bar{W}} \le \|v\|_W \le l_\alpha \|v\|_{\bar{W}} \quad \text{for } v \in \mathbb{B}_W, \tag{4.7}$$

where

$$l_\alpha := 1 + e = 1 + \frac{\alpha d}{\gamma_\alpha - \alpha}. \tag{4.8}$$

Combining these facts, from Theorem 2.6 we have the following result.

Theorem 4.4. *Suppose that Assumption 4.1 holds and $\theta \in \mathcal{M}(S)$. Then, for each $\alpha \in (0,1)$:*

(a) The discounted payoff game has a value $V_\alpha^\theta \in \mathbb{C}_W$ and

$$\left\| V_\alpha^\theta \right\|_W \le \frac{M}{1-\alpha}.$$

(b) The value V_α^θ satisfies $T_{(\alpha,\theta)} V_\alpha^\theta = V_\alpha^\theta$, and there exists $(\varphi_*^1, \varphi_*^2) \in \Phi^1 \times \Phi^2$ such that, for all $x \in X$, $\varphi_*^1(x) \in \mathbb{A}(x)$ and $\varphi_*^2(x) \in \mathbb{B}(x)$ satisfy

$$V_\alpha^\theta(x) = r(x, \varphi_*^1, \varphi_*^2) + \alpha \int_S V_\alpha^\theta[F(x, \varphi_*^1, \varphi_*^2, s)]\theta(ds)$$

$$= \max_{\varphi^1 \in \mathbb{A}(x)} \left[r(x, \varphi^1, \varphi_*^2) + \alpha \int_S V_\alpha^\theta[F(x, \varphi^1, \varphi_*^2, s)]\theta(ds) \right] \tag{4.9}$$

$$= \min_{\varphi^2 \in \mathbb{B}(x)} \left[r(x, \varphi_*^1, \varphi^2) + \alpha \int_S V_\alpha^\theta[F(x, \varphi_*^1, \varphi^2, s)]\theta(ds) \right]. \tag{4.10}$$

In addition, $\pi_*^1 = \{\varphi_*^1\} \in \Pi_s^1$ and $\pi_*^2 = \{\varphi_*^2\} \in \Pi_s^2$ form an optimal pair of strategies.

In addition, Theorem 3.3 yields the following result related to the average payoff game.

Theorem 4.5. *Under Assumptions 4.1 and 4.2, the average payoff game has a value* $J(\cdot) = j^*$, *that is,*

$$j^* = \inf_{\pi^2 \in \Pi^2} \sup_{\pi^1 \in \Pi^1} J(x, \pi^1, \pi^2) = \sup_{\pi^1 \in \Pi^1} \inf_{\pi^2 \in \Pi^2} J(x, \pi^1, \pi^2) \quad \forall x \in X.$$

Further, both players have optimal strategies.

Remark 4.2. The average criterion is analyzed according to the VDFA introduced in Sect. 3.2; for ease reference we formulate it again in terms of the distribution θ. Let $z \in X$ be a fixed state, and define, for $\alpha \in (0,1)$ and $x \in X$,

$$j_\alpha^\theta := (1-\alpha)V_\alpha^\theta(z), \quad h_\alpha^\theta(x) := V_\alpha^\theta(x) - V_\alpha^\theta(z). \tag{4.11}$$

Observe that, from Theorem 4.4(b), for all $x \in X$ we have

$$j_\alpha^\theta + h_\alpha^\theta(x) = T_{(\alpha,\theta)}h_\alpha^\theta(x) = r(x, \varphi_*^1, \varphi_*^2) + \alpha \int_S h_\alpha^\theta[F(x, \varphi_*^1, \varphi_*^2, s)]\theta(ds)$$

$$= \max_{\varphi^1 \in \mathbb{A}(x)} \left[r(x, \varphi^1, \varphi_*^2) + \alpha \int_S h_\alpha^\theta[F(x, \varphi^1, \varphi_*^2, s)]\theta(ds) \right]$$

$$= \min_{\varphi^2 \in \mathbb{B}(x)} \left[r(x, \varphi_*^1, \varphi^2) + \alpha \int_S h_\alpha^\theta[F(x, \varphi_*^1, \varphi^2, s)]\theta(ds) \right]. \tag{4.12}$$

Then, from [42, Theorem 4.3], under Assumptions 4.1 and 4.2 (see (3.9)), we have

$$\lim_{t \to \infty} j_{\alpha_t}^\theta = j^*, \tag{4.13}$$

for any sequence $\{\alpha_t\}$ of discount factors such that $\alpha_t \nearrow 1$. Moreover (see (3.25)),

$$\sup_{\alpha \in (0,1)} \left\| h_\alpha^\theta \right\|_W < \infty. \tag{4.14}$$

4.2 The Discounted Empirical Game

Let $\theta_t \in \mathbb{P}(S)$, for $t \in \mathbb{N}_0$, be the empirical distribution of the disturbance process $\{\xi_t\}$ (see Appendix B.1). That is, for a given probability measure $\nu \in \mathbb{P}(S)$,

$$\theta_0 := \nu,$$

$$\theta_t(D) = \theta_t(D)(\omega) := \frac{1}{t} \sum_{i=0}^{t-1} 1_D(\xi_i(\omega)) \quad \forall t \in \mathbb{N}, D \in \mathscr{B}(S), \omega \in \Omega.$$

Note that for each $D \in B(S)$, $\theta_t(D)(\cdot)$ is a random variable, and for each $\omega \in \Omega$, $\theta_t(\cdot)(\omega)$ is the uniform distribution on the set $\{\xi_0(\omega), \ldots, \xi_{t-1}(\omega)\} \subset S$.

We consider the approximating zero-sum Markov game model of the form:

$$\mathscr{G}\mathscr{M}_t^{\alpha} := (X, A, B, \mathbb{K}_A, \mathbb{K}_B, Q_t, r), \tag{4.15}$$

where, for all $D \in \mathscr{B}(X)$ and $(x, a, b) \in \mathbb{K}$,

$$Q_t(D|x, a, b) = \int_S 1_D[F(x, a, b, s)]\theta_t(ds)$$

$$= \frac{1}{t}\sum_{i=1}^{t} 1_D[F(x, a, b, \xi_i)].$$

The empirical approximation scheme consists in solving the approximate game $\mathscr{G}\mathscr{M}_t$, for each $t \in \mathbb{N}$. That is, the discounted game is analyzed when both players use the empirical distribution θ_t instead of the original distribution θ. This procedure leads to an optimal pair of strategies $(\pi_t^1, \pi_t^2) \in \Pi_s^1 \times \Pi_s^2$ for the game $\mathscr{G}\mathscr{M}_t^{\alpha}$, for each $t \in \mathbb{N}$, provided, of course, that the corresponding value of the game $V_\alpha^{\theta_t}$ exists. Under this setting, our hope is that the optimal pair (π_t^1, π_t^2) will have a good performance in the game $\mathscr{G}\mathscr{M}$, whenever the sequence of empirical values $\left\{V_\alpha^{\theta_t}\right\}$ gives a good approximation to the value V_α^{θ}. We now introduce these ideas in precise terms as follows.

For each $t \in \mathbb{N}_0$, let

$$V_\alpha^{\theta_t}(x, \pi^1, \pi^2) := E_t^{x, \pi^1, \pi^2}\left[\sum_{i=0}^{\infty} \alpha^i r(x_i, a_i, b_i)\right], \tag{4.16}$$

be the α-discounted expected payoff function in which all random variables $\xi_0^t, \xi_1^t, \ldots$ have the same distribution θ_t.

Observe that, under Assumption 4.3, θ_t is in $\mathscr{M}(S)$ for every $t \in \mathbb{N}$, that is,

$$\int_S W[F(x, a, b, s)]\theta_t(ds)(\omega) \leq \beta W(x) + d \quad \forall (x, a, b) \in \mathbb{K}.$$

Then Theorem 4.4 yields the following result.

Theorem 4.6. *Suppose that Assumptions 4.1 and 4.3 hold. Then for each $t \in \mathbb{N}$ and $\omega \in \Omega$,*
(a) the game $\mathscr{G}\mathscr{M}_t^{\alpha}$ has a value $V_\alpha^{\theta_t} = V_\alpha^{\theta_t}(\omega) \in \mathbb{C}_W$ such that

$$\left\|V_\alpha^{\theta_t}\right\|_W \leq \frac{M}{1-\alpha} \quad and \quad T_{(\alpha, \theta_t)}V_\alpha^{\theta_t} = V_\alpha^{\theta_t};$$

(b) *there exists* $(\varphi_t^1, \varphi_t^2) = (\varphi_t^1(\omega), \varphi_t^2(\omega)) \in \Phi^1 \times \Phi^2$ *such that, for all* $x \in X$,
$\varphi_t^1(x, \omega) := \varphi_t^1(\cdot|x, \omega) \in \mathbb{A}(x)$ *and* $\varphi_t^2(x, \omega) := \varphi_t^2(\cdot|x, \omega) \in \mathbb{B}(x)$ *satisfy*

$$V_\alpha^{\theta_t}(x) = r(x, \varphi_t^1, \varphi_t^2) + \alpha \int_S V_\alpha^{\theta_t}[F(x, \varphi_t^1, \varphi_t^2, s)]\theta_t(ds)$$

$$= \max_{\varphi^1 \in \mathbb{A}(x)} \left[r(x, \varphi^1, \varphi_t^2) + \alpha \int_S V_\alpha^{\theta_t}[F(x, \varphi^1, \varphi_t^2, s)]\theta_t(ds) \right] \quad (4.17)$$

$$= \min_{\varphi^2 \in \mathbb{B}(x)} \left[r(x, \varphi_t^1, \varphi^2) + \alpha \int_S V_\alpha^{\theta_t}[F(x, \varphi_t^1, \varphi^2, s)]\theta_t(ds) \right]. \quad (4.18)$$

Remark 4.3. Observe that, for each $t \in \mathbb{N}_0$, the value function $V_\alpha^{\theta_t}$ is a random function, and $\varphi_t^i(x, \omega)$, for $i = 1, 2$, define a random optimal pair of strategies $(\pi_t^1, \pi_t^2) := (\{\varphi_t^1\}, \{\varphi_t^2\}) \in \Pi_s^1 \times \Pi_s^2$ for the game $\mathscr{GM}_t^\alpha$.

4.2.1 Empirical Estimation Process

The key points to obtain the approximation of the empirical values $V_\alpha^{\theta_t}$ to the value V_α^θ are the convergence properties of the empirical distribution. At first glance, from Proposition B.5, in Appendix B, we have that θ_t converges weakly to θ a.s. Thus, for each $(x, a, b) \in \mathbb{K}$ and continuous and bounded function u on X, (see Definition B.1)

$$\int_S u(F(x, a, b, s))\theta_t(ds) \to \int_S u(F(x, a, b, s))\theta(ds) \text{ a.s., as } t \to \infty.$$

However, in the scenario of possibly unbounded payoff and arbitrary disturbance space S, this class of convergence is not sufficient for our objectives. In fact, we need uniform convergence on the set $\mathbb{K}$. In order to state our estimation process, we impose the following assumption.

Assumption 4.7 *The family of functions*

$$\mathscr{V}_W := \left\{ \frac{V_\alpha^\theta(F(x, a, b, \cdot))}{W(x)} : (x, a, b) \in \mathbb{K} \right\} \quad (4.19)$$

is equicontinuous on S.

Proposition 4.1. *Under Assumption 4.7,*

$$\Delta_t \to 0 \text{ a.s., as } t \to \infty, \quad (4.20)$$

where

$$\Delta_t := \sup_{(x,a,b)\in\mathbb{K}} \left| \int_S \frac{V_\alpha^\theta(F(x,a,b,s))}{W(x)} \theta_t(ds) - \int_S \frac{V_\alpha^\theta(F(x,a,b,s))}{W(x)} \theta(ds) \right|.$$

Proof. Observe that from Theorem 4.4(a), the family of functions $\mathscr{V}_W$ is uniformly bounded. Then, under Assumption 4.7, the relation (4.20) follows from Proposition B.6 in Appendix B. ∎

4.2.2 Discounted Optimal Strategies

Let us fix $(x,\omega) \in X \times \Omega$, and consider the multifunction given by $(x,\omega) \longmapsto \mathbb{A}(x)$. Since $A(x)$ is a compact subset of A, $\mathbb{A}(x)$ is a compact subset of $\mathbb{P}(A)$ (with the weak topology). Consider the optimal pair of strategies $(\pi_t^1, \pi_t^2) = (\{\varphi_t^1\}, \{\varphi_t^2\}) \in \Pi_s^1 \times \Pi_s^2$ for the game model $\mathscr{GM}_t^\alpha$ (see Remark 4.3). Then, from Proposition A.4 in Appendix A, there exists $\varphi_\infty^1 \in \Phi^1$ such that $\varphi_\infty^1(x,\omega) = \varphi_\infty^1(\cdot|x,\omega) \in \mathbb{A}(x)$ is an accumulation point of $\{\varphi_t^1(x,\omega)\}$. Similarly, there exists $\varphi_\infty^2 \in \Phi^2$ such that $\varphi_\infty^2(x,\omega) = \varphi_\infty^2(\cdot|x,\omega) \in \mathbb{B}(x)$ is an accumulation point of $\{\varphi_t^2(x,\omega)\}$.

Define the strategies $\pi_\infty^1 = \{\varphi_\infty^1\} \in \Pi_s^1$ and $\pi_\infty^2 = \{\varphi_\infty^2\} \in \Pi_s^2$. Hence, we can state our results related to the discounted empirical approximation as follows.

Theorem 4.8. *Under Assumptions 4.1, 4.3, and 4.7, $P - a.s.$*

(a) $\left\| V_\alpha^{\theta_t} - V_\alpha^\theta \right\|_W \to 0$ *as* $t \to \infty$;

(b) the random pair of strategies $(\pi_\infty^1, \pi_\infty^2) \in \Pi_s^1 \times \Pi_s^2$ is optimal for the game $\mathscr{GM}$.

Furthermore,
(c) there exist an optimal (nonrandom) pair of strategies $(\hat{\pi}_\infty^1, \hat{\pi}_\infty^2) \in \Pi_s^1 \times \Pi_s^2$ for the game $\mathscr{GM}$ defined as $\hat{\pi}_\infty^i = \{\hat{\varphi}_\infty^i\}$ where

$$\hat{\varphi}_\infty^i(\cdot|x) = \int_\Omega \varphi_t^i(\cdot|x,\omega)P(d\omega), \quad i = 1,2.$$

Proof. (a) Since $\theta_t \in \mathscr{M}(S)$, for $t \in \mathbb{N}_0$, from (4.6) we have that the operator $T_{(\alpha,\theta_t)}$ is a contraction. Hence, from Theorems 4.4 and 4.6, for each $t \in \mathbb{N}_0$,

$$\left\| V_\alpha^\theta - V_\alpha^{\theta_t} \right\|_{\bar{W}} \leq \left\| T_{(\alpha,\theta)}V_\alpha^\theta - T_{(\alpha,\theta_t)}V_\alpha^\theta \right\|_{\bar{W}} + \left\| T_{(\alpha,\theta_t)}V_\alpha^\theta - T_{(\alpha,\theta_t)}V_\alpha^{\theta_t} \right\|_{\bar{W}}$$

$$\leq \left\| T_{(\alpha,\theta)}V_\alpha^\theta - T_{(\alpha,\theta_t)}V_\alpha^\theta \right\|_{\bar{W}} + \gamma_\alpha \left\| V_\alpha^\theta - V_\alpha^{\theta_t} \right\|_{\bar{W}} \quad \text{a.s.}$$

Thus

$$\left\| V_\alpha^{\theta_t} - V_\alpha^\theta \right\|_{\bar{W}} \leq \frac{1}{1-\gamma_\alpha} \left\| T_{(\alpha,\theta)}V_\alpha^\theta - T_{(\alpha,\theta_t)}V_\alpha^\theta \right\|_{\bar{W}}. \tag{4.21}$$

On the other hand, using the fact that $\bar{W}(\cdot) > W(\cdot)$, for each $x \in X$ and $t \in \mathbb{N}_0$,

$$\left\| T_{(\alpha,\theta)} V_\alpha^\theta - T_{(\alpha,\theta_t)} V_\alpha^\theta \right\|_{\bar{W}}$$

$$\leq \sup_{x \in X} \sup_{\varphi^1 \in \mathbb{A}(x), \varphi^2 \in \mathbb{B}(x)} \left| \int_S \frac{V_\alpha^\theta [F(x, \varphi^1, \varphi^2, s)]}{W(x)} \theta(ds) - \int_S \frac{V_\alpha^\theta [F(x, \varphi^1, \varphi^2, s)]}{W(x)} \theta_t(ds) \right|$$

$$= \Delta_t. \tag{4.22}$$

Combining (4.21) and (4.22) we get

$$\left\| V_\alpha^{\theta_t} - V_\alpha^\theta \right\|_{\bar{W}} \leq \frac{1}{1 - \gamma_\alpha} \Delta_t,$$

and from (4.7)

$$\left\| V_\alpha^{\theta_t} - V_\alpha^\theta \right\|_W \leq \frac{l_\alpha}{1 - \gamma_\alpha} \Delta_t. \tag{4.23}$$

Thus, (4.20) yields part (a).

(b) Since for each $(x, \omega) \in X \times \Omega$, $\varphi_\infty^1(x, \omega) = \varphi_\infty^1(\cdot | x, \omega) \in \mathbb{A}(x)$ is an accumulation point of $\{\varphi_t^1(x, \omega)\}$, there exists a subsequence $\{\varphi_{t_k}^1(x, \omega)\}$ of $\{\varphi_t^1(x, \omega)\}$ such that $\varphi_\infty^1(x, \omega) = \lim_{k \to \infty} \varphi_{t_k}^1(x, \omega)$. Under similar arguments, there exists a subsequence $\{\varphi_{t_k}^2(x, \omega)\}$ of $\{\varphi_t^2(\cdot | x, \omega)\}$ such that $\varphi_\infty^2(x, \omega) = \lim_{k \to \infty} \varphi_{t_k}^2(x, \omega)$. Observe that we can use the same subsequence $\{t_k\}$ for both cases. In the remainder of the proof, to ease notation, we let $t_k = k$.

We shall now proceed to prove the optimality of the pair $(\pi_\infty^1, \pi_\infty^2) \in \Pi_s^1 \times \Pi_s^2$.

Firstly, observe that, for each $x \in X$, as $t \to \infty$,

$$\sup_{(a,b) \in A(x) \times B(x)} \left| \int_S V_\alpha^{\theta_t}(F(x,a,b,s)) \theta_t(ds) - \int_S V_\alpha^\theta(F(x,a,b,s)) \theta(ds) \right| \to 0 \quad \text{a.s.} \tag{4.24}$$

Indeed,

$$\left| \int_S V_\alpha^{\theta_t}(F(x,a,b,s)) \theta_t(ds) - \int_S V_\alpha^\theta(F(x,a,b,s)) \theta(ds) \right|$$

$$\leq \int_S \left| V_\alpha^{\theta_t}(F(x,a,b,s)) - V_\alpha^\theta(F(x,a,b,s)) \right| \theta_t(ds)$$

$$+ \left| \int_S V_\alpha^\theta(F(x,a,b,s)) \theta_t(ds) - \int_S V_\alpha^\theta(F(x,a,b,s)) \theta(ds) \right|$$

$$\leq \left\| V_\alpha^{\theta_t} - V_\alpha^\theta \right\|_W (\beta W(x) + d) + \Delta_t W(x).$$

Thus, (4.24) follows from part (a) and (4.20).

Now, from (4.17),

$$V_\alpha^{\theta_k}(x) = \max_{\varphi^1 \in \mathbb{A}(x)} \left[r(x, \varphi^1, \varphi_k^2) + \alpha \int_S V_\alpha^{\theta_k}[F(x, \varphi^1, \varphi_k^2, s)] \theta_k(ds) \right]. \tag{4.25}$$

In addition, for any fixed $\bar{\varphi}^1 \in \mathbb{A}(x)$

$$\liminf_k \max_{\varphi^1 \in \mathbb{A}(x)} \left[r(x, \varphi^1, \varphi_k^2) + \alpha \int_S V_\alpha^{\theta_k}[F(x, \varphi^1, \varphi_k^2, s)] \theta_k(ds) \right]$$

$$\geq \liminf_k \left[r(x, \bar{\varphi}^1, \varphi_k^2) + \alpha \int_S V_\alpha^{\theta_k}[F(x, \bar{\varphi}^1, \varphi_k^2, s)] \theta_k(ds) \right]. \tag{4.26}$$

On the other hand,

$$\int_S V_\alpha^{\theta_k}[F(x, \bar{\varphi}^1, \varphi_k^2, s)] \theta_k(ds) = \int_S V_\alpha^{\theta_k}[F(x, \bar{\varphi}^1, \varphi_k^2, s)] \theta_k(ds)$$

$$- \int_S V_\alpha^{\theta}[F(x, \bar{\varphi}^1, \varphi_k^2, s)] \theta(ds) + \int_S V_\alpha^{\theta}[F(x, \bar{\varphi}^1, \varphi_k^2, s)] \theta(ds).$$

Then, from (4.24), Fatou's Lemma, and using the continuity of the functions V_α^θ and F,

$$\liminf_k \int_S V_\alpha^{\theta_k}[F(x, \bar{\varphi}^1, \varphi_k^2, s)] \theta_k(ds) = \liminf_k \int_S V_\alpha^{\theta}[F(x, \bar{\varphi}^1, \varphi_k^2, s)] \theta(ds)$$

$$\geq \int_S V_\alpha^{\theta}[F(x, \bar{\varphi}^1, \varphi_\infty^2, s)] \theta(ds) \quad a.s. \tag{4.27}$$

Therefore, taking liminf as $k \to \infty$ in (4.25), the relations (4.26) and (4.27) together with part (a) yield

$$V_\alpha^\theta(x) \geq r(x, \bar{\varphi}^1, \varphi_\infty^2) + \alpha \int_S V_\alpha^\theta[F(x, \bar{\varphi}^1, \varphi_\infty^2, s)] \theta(ds).$$

Since $\bar{\varphi}^1 \in \mathbb{A}(x)$ is arbitrary, we have

$$V_\alpha^\theta(x) \geq \max_{\varphi^1 \in \mathbb{A}(x)} \left[r(x, \varphi^1, \varphi_\infty^2) + \alpha \int_S V_\alpha^\theta[F(x, \varphi^1, \varphi_\infty^2, s)] \theta(ds) \right].$$

This implies

$$V_\alpha^\theta(x) = \max_{\varphi^1 \in \mathbb{A}(x)} \left[r(x, \varphi^1, \varphi_\infty^2) + \alpha \int_S V_\alpha^\theta[F(x, \varphi^1, \varphi_\infty^2, s)] \theta(ds) \right], \tag{4.28}$$

because (see (4.9))

$$V_\alpha^\theta(x) = \min_{\varphi^2 \in \mathbb{B}(x)} \max_{\varphi^1 \in \mathbb{A}(x)} \left[r(x, \varphi^1, \varphi^2) + \alpha \int_S V_\alpha^\theta [F(x, \varphi^1, \varphi^2, s)] \theta(ds) \right]$$

$$\leq \max_{\varphi^1 \in \mathbb{A}(x)} \left[r(x, \varphi^1, \varphi^2) + \alpha \int_S V_\alpha^\theta [F(x, \varphi^1, \varphi^2, s)] \theta(ds) \right], \quad \forall \varphi^2 \in \mathbb{B}(x).$$

Similarly, from (4.18),

$$V_\alpha^{\theta_k}(x) = \min_{\varphi^2 \in \mathbb{B}(x)} \left[r(x, \varphi_k^1, \varphi^2) + \alpha \int_S V_\alpha^{\theta_k} [F(x, \varphi_k^1, \varphi^2, s)] \theta_k(ds) \right],$$

and for an arbitrary and fixed $\bar\varphi^2 \in \mathbb{B}(x)$

$$\limsup_k \min_{\varphi^2 \in \mathbb{B}(x)} \left[r(x, \varphi_k^1, \varphi^2) + \alpha \int_S V_\alpha^{\theta_k} [F(x, \varphi_k^1, \varphi^2, s)] \theta_k(ds) \right]$$

$$\leq \limsup_k \left[r(x, \varphi_k^1, \bar\varphi^2) + \alpha \int_S V_\alpha^{\theta_k} [F(x, \varphi_k^1, \bar\varphi^2, s)] \theta_k(ds) \right].$$

Thus, applying Fatou's Lemma with limsup, we obtain

$$V_\alpha^\theta(x) \leq r(x, \varphi_\infty^1, \bar\varphi^2) + \alpha \int_S V_\alpha^\theta [F(x, \varphi_\infty^1, \bar\varphi^2, s)] \theta(ds),$$

which, in turn, implies

$$V_\alpha^\theta(x) = \min_{\varphi^2 \in \mathbb{B}(x)} \left[r(x, \varphi_\infty^1, \varphi^2) + \alpha \int_S V_\alpha^\theta [F(x, \varphi_\infty^1, \varphi^2, s)] \theta(ds) \right]. \tag{4.29}$$

Finally, combining (4.28) and (4.29), and applying standard procedures in game theory, we prove that $(\pi_\infty^1, \pi_\infty^2) \in \Pi_s^1 \times \Pi_s^2$ is a random optimal pair of strategies for the game $\mathscr{GM}$.

(c) We define

$$H(x, a, b) := r(x, a, b) + \alpha \int_S V_\alpha^\theta [F(x, a, b, s)] \theta(ds), \quad (x, a, b) \in \mathbb{K}. \tag{4.30}$$

Observe that, from (4.29) and (1.5),

$$V_\alpha^\theta(x) = \min_{\varphi^2 \in \mathbb{B}(x)} H(x, \varphi_\infty^1(\omega), \varphi^2)$$

$$= \min_{\varphi^2 \in \mathbb{B}(x)} \int_{A(x)} H(x, a, \varphi^2) \varphi_\infty^1(da|x, \omega) \quad a.s., \quad x \in X.$$

Hence,

$$V_\alpha^\theta(x) = \int_\Omega \min_{\varphi^2 \in \mathbb{B}(x)} \int_{A(x)} H(x, a, \varphi^2) \varphi_\infty^1(da|x, \omega) P(d\omega)$$

$$\leq \min_{\varphi^2 \in \mathbb{B}(x)} \int_{A(x)} H(x, a, \varphi^2) \int_\Omega \varphi_\infty^1(da|x, \omega) P(d\omega)$$

$$= \min_{\varphi^2 \in \mathbb{B}(x)} \int_{A(x)} H(x, a, \varphi^2) \hat{\varphi}_\infty^1(da|x)$$

$$= \min_{\varphi^2 \in \mathbb{B}(x)} H(x, \hat{\varphi}_\infty^1, \varphi^2), \quad x \in X.$$

Therefore, from (4.30) and Theorem 4.4, for all $x \in X$,

$$V_\alpha^\theta(x) = \min_{\varphi^2 \in \mathbb{B}(x)} \left[r(x, \hat{\varphi}_\infty^1, \varphi^2) + \alpha \int_S V_\alpha^\theta [F(x, \hat{\varphi}_\infty^1, \varphi^2, s)] \theta(ds) \right]. \qquad (4.31)$$

Similarly, we can prove that, for each $x \in X$,

$$V_\alpha^\theta(x) = \max_{\varphi^1 \in \mathbb{A}(x)} \left[r(x, \varphi^1, \hat{\varphi}_\infty^2) + \alpha \int_S V_\alpha^\theta [F(x, \varphi^1, \hat{\varphi}_\infty^2, s)] \theta(ds) \right],$$

which, combined with (4.31), yields the optimality of the pair $(\hat{\pi}_\infty^1, \hat{\pi}_\infty^2) \in \Pi_s^1 \times \Pi_s^2$ for the game $\mathcal{GM}$. ∎

4.3 Empirical Approximation Under Average Criterion

The empirical approximation scheme for the average criterion is obtained by combining the VDFA and a suitable convergence property of the empirical process, similar to the procedure followed in Sect. 3.3. Therefore, we will take advantage of the results introduced in previous sections for the discounted criterion. However, due to the additional difficulties in the asymptotic analysis of the average payoff, the following stronger condition is needed.

Assumption 4.9 *(a) The disturbance space S is the k-dimensional Euclidean space* $\mathfrak{R}^k$.
(b) Let $m > \max\{2, k\}$ *be an arbitrary real number and* $\bar{m} := km/[(m-k)(m-2)]$. *Then* $E|\xi_0|^{\bar{m}} < \infty$.
(c) The family of functions (see (4.11) and (4.19))

$$\bar{\mathcal{V}}_W := \left\{ \frac{h_\alpha^\theta(F(x, a, b, \cdot))}{W(x)} : (x, a, b) \in \mathbb{K}, \alpha \in (0, 1) \right\},$$

or equivalently

$$\hat{\mathcal{V}}_W := \left\{ \frac{V_\alpha^\theta \left(F(x,a,b,.) \right)}{W(x)} : (x,a,b) \in \mathbb{K}, \alpha \in (0,1) \right\},$$

is equi-Lipschitzian on $\mathfrak{R}^k$. *That is, there exists a constant* $L_h > 0$ *such that, for every* $s, s' \in \mathfrak{R}^k$ *and* $(x,a,b) \in \mathbb{K}$,

$$\left| \frac{h_\alpha^\theta \left(F(x,a,b,s) \right)}{W(x)} - \frac{h_\alpha^\theta \left(F(x,a,b,s') \right)}{W(x)} \right| \leq L_h \left| s - s' \right|,$$

where $|\cdot|$ *is the corresponding Euclidean distance in* $\mathfrak{R}^k$.

Remark 4.4 (Equicontinuity and Equi-Lipschitz Conditions). Clearly, in the case $S = \mathfrak{R}^k$, the equi-Lipschitz Assumption 4.9(c) implies the equicontinuity Assumption 4.7.

Proposition 4.2. *Under Assumptions 4.1, 4.2, and 4.9, there exists a constant* $\bar{M}$ *such that*

$$E \left[\bar{\Delta}_t \right] \leq \bar{M} t^{-1/m}, \tag{4.32}$$

where

$$\bar{\Delta}_t := \sup_{(x,a,b) \in \mathbb{K}, \alpha \in (0,1)} \left| \int_{\mathfrak{R}^k} \frac{h_\alpha \left(F(x,a,b,s) \right)}{W(x)} \theta_t(ds) - \int_{\mathfrak{R}^k} \frac{h_\alpha \left(F(x,a,b,s) \right)}{W(x)} \theta(ds) \right|. \tag{4.33}$$

Proof. From (4.14), the family of functions $\hat{\mathcal{V}}_W$ is uniformly bounded. Thus, by applying Proposition B.7 in Appendix B we prove (4.32). ∎

In order to introduce the empirical VDFA, we follow similar ideas as in Sect. 3.3.1. Let $v \in (0, 1/2m)$ be an arbitrary real number where m is the constant introduced in Assumption 4.9(b). We fix an arbitrary nondecreasing sequence of discount factors $\{\bar{\alpha}_t\}$ such that

$\bar{F}.1$ $(1 - \bar{\alpha}_t)^{-1} = O(t^v)$ as $t \to \infty$;

$\bar{F}.2$ $\lim_{n \to \infty} \dfrac{\kappa(n)}{n} = 0$,

where $\kappa(n)$ is the number of changes of value of $\{\bar{\alpha}_t\}$ among the first n terms (see Conditions F.1 and F.2 and the succeeding example in Sect. 3.3.1).

For a fixed $t \in \mathbb{N}_0$, let $V_{\bar{\alpha}_t}^{\theta_t}(\cdot, \cdot, \cdot)$ be the $\bar{\alpha}_t$-discounted payoff function under the empirical distribution θ_t (see (4.16)), and we denote by $V_{\bar{\alpha}_t}^{\theta_t}(\cdot)$ the corresponding value of the game $\mathscr{GM}_t^{\bar{\alpha}_t}$ (see (4.15), Theorems 4.4 and 4.6). The functions

$h_{\bar{\alpha}_t}^{\theta_t}(\cdot)$ and $j_{\bar{\alpha}_t}^{\theta_t}$ are defined accordingly (see (4.11)). Hence, from Theorem 4.4(b) (see (4.12)), there exists a random pair $(\bar{\varphi}_t^1, \bar{\varphi}_t^2) \in \Phi^1 \times \Phi^2$ such that, for every $x \in X$,

$$j_{\bar{\alpha}_t}^{\theta_t} + h_{\bar{\alpha}_t}^{\theta_t}(x) = T_{(\bar{\alpha}_t, \theta_t)} h_{\bar{\alpha}_t}^{\theta_t}(x)$$

$$= r(x, \bar{\varphi}_t^1, \bar{\varphi}_t^2) + \bar{\alpha}_t \int_S h_{\bar{\alpha}_t}^{\theta_t}[F(x, \bar{\varphi}_t^1, \bar{\varphi}_t^2, s)] \theta_t(ds)$$

$$= \max_{\varphi^1 \in \mathbb{A}(x)} \left[r(x, \varphi^1, \bar{\varphi}_t^2) + \bar{\alpha}_t \int_S h_{\bar{\alpha}_t}^{\theta_t}[F(x, \varphi^1, \bar{\varphi}_t^2, s)] \theta_t(ds) \right]$$

$$= \min_{\varphi^2 \in \mathbb{B}(x)} \left[r(x, \bar{\varphi}_t^1, \varphi^2) + \bar{\alpha}_t \int_S h_{\bar{\alpha}_t}^{\theta_t}[F(x, \bar{\varphi}_t^1, \varphi^2, s)] \theta_t(ds) \right]. \quad (4.34)$$

For each $t \in \mathbb{N}_0$ (see (4.4)–(4.8)) we define

$$\gamma_t \equiv \gamma_{\bar{\alpha}_t} := \frac{1 + \bar{\alpha}_t}{2} \in (\bar{\alpha}_t, 1),$$

$$e_t := d \left(\frac{\gamma_t}{\bar{\alpha}_t} - 1 \right)^{-1} = d \left(\frac{2\bar{\alpha}_t}{1 - \bar{\alpha}_t} \right),$$

and

$$l_t \equiv l_{\bar{\alpha}_t} := 1 + e_t = 1 + \frac{2d\bar{\alpha}_t}{1 - \bar{\alpha}_t}.$$

It is easy to see that

$$\frac{l_t}{1 - \gamma_t} \leq 2(1 + d)(1 - \bar{\alpha}_t)^{-2},$$

which, from Condition $\bar{F}.1$, yields

$$\frac{l_t}{1 - \gamma_t} = O(t^{2\nu}) \text{ as } t \to \infty. \quad (4.35)$$

Moreover, applying similar arguments as in the proof of Theorem 4.8 (see (4.23)) and from definition of the function h_α^θ (see (4.11)), we can obtain

$$\left\| V_{\bar{\alpha}_t}^{\theta_t} - V_{\bar{\alpha}_t}^\theta \right\|_W \leq \frac{l_t}{1 - \gamma_t} \bar{\Delta}_t.$$

Hence, for all $(\pi^1, \pi^2) \in \Pi^1 \times \Pi^2$ and $x \in X$, from (4.32) and (4.35),

$$E_x^{\pi^1, \pi^2} \left\| V_{\bar{\alpha}_t}^{\theta_t} - V_{\bar{\alpha}_t}^\theta \right\|_W = O(t^{2\nu}) O(t^{-1/m}), \text{ as } t \to \infty. \quad (4.36)$$

Then, because $2\nu < 1/m$, we get

$$\lim_{t\to\infty} E_x^{\pi^1,\pi^2} \left\| V_{\bar{\alpha}_t}^{\theta_t} - V_{\bar{\alpha}_t}^{\theta} \right\|_W = 0. \tag{4.37}$$

Again, from definition of the functions $h_\alpha^\theta(x)$ and j_α^θ (see (4.11)), we have

$$\lim_{t\to\infty} E_x^{\pi^1,\pi^2} \left\| h_{\bar{\alpha}_t}^{\theta_t} - h_{\bar{\alpha}_t}^{\theta} \right\|_W = 0 \tag{4.38}$$

and

$$\lim_{t\to\infty} E_x^{\pi^1,\pi^2} \left| j_{\bar{\alpha}_t}^{\theta_t} - j_{\bar{\alpha}_t}^{\theta} \right| = 0. \tag{4.39}$$

On the other hand, following similar ideas as in the proofs of Lemma 3.2(b) and relation (2.50), and with the necessary changes, we obtain

$$\lim_{t\to\infty} E_x^{\pi^1,\pi^2} \left\| h_{\bar{\alpha}_t}^{\theta_t} - h_{\bar{\alpha}_t}^{\theta} \right\|_W W(x_t) = 0 \tag{4.40}$$

and

$$\lim_{t\to\infty} E_x^{\pi^1,\pi^2} \bar{\Delta}_t W(x_t) = 0. \tag{4.41}$$

4.4 Average Optimal Strategies

Let $(\pi_*^1, \pi_*^2) \in \Pi^1 \times \Pi^2$ be the pair of strategies determined by $(\bar{\varphi}_t^1, \bar{\varphi}_t^2) \in \Phi^1 \times \Phi^2$ (see (4.34)). That is, $\pi_*^i = \{\bar{\varphi}_t^i\} = \{\bar{\varphi}_t^i(\cdot|x, \omega)\}$ for $i = 1, 2$. In addition, let $\hat{\pi}_*^i = \{\hat{\varphi}_t^i\}$, $i = 1, 2$, be the strategies defined by

$$\hat{\varphi}_t^i(\cdot|x) = \int_\Omega \bar{\varphi}_t^i(\cdot|x, \omega) P(d\omega).$$

Then, our main result is stated as follows.

Theorem 4.10. *Under Assumptions 4.1, 4.2, and 4.9, the pair $(\pi_*^1, \pi_*^2) \in \Pi^1 \times \Pi^2$ is a random pair of average optimal strategies for the game $\mathcal{GM}$, that is,*

$$j^* = \inf_{\pi^2 \in \Pi^2} J(x, \pi_*^1, \pi^2) = \sup_{\pi^1 \in \Pi^1} J(x, \pi^1, \pi_*^2) \quad \forall x \in X. \tag{4.42}$$

Furthermore, the strategies $\hat{\pi}_*^i = \{\hat{\varphi}_t^i\}$, for $i = 1, 2$, form an average optimal pair of nonrandom strategies.

Proof. We first prove the optimality of $\pi_*^2 = \{\bar{\varphi}^2(\cdot|x_t, \omega)\} = \{\bar{\varphi}_t^2\}$, for which we will show

$$j^* = \sup_{\pi^1 \in \Pi^1} J(x, \pi^1, \pi_*^2) \quad \forall x \in X.$$

Let $\pi^1 = \{\pi_t^1\} \in \Pi^1$ be an arbitrary strategy for player 1. Then

$$\mathcal{L}_t := r(x_t, \pi_t^1, \bar{\varphi}_t^2) + \bar{\alpha}_t \int_{\Re^k} h_{\bar{\alpha}_t}^\theta [F(x_t, \pi_t^1, \bar{\varphi}_t^2, s)]\theta(ds) - j_{\bar{\alpha}_t}^\theta - h_{\bar{\alpha}_t}^\theta(x_t)$$

$$= r(x_t, \pi_t^1, \bar{\varphi}_t^2) + \bar{\alpha}_t E_x^{\pi^1, \pi_*^2} \left[h_{\bar{\alpha}_t}^\theta(x_{t+1}) \mid h_t \right] - j_{\bar{\alpha}_t}^\theta - h_{\bar{\alpha}_t}^\theta(x_t),$$

which implies

$$n^{-1} E_x^{\pi^1, \pi_*^2} \left[\sum_{t=0}^{n-1} \left(r(x_t, a_t, b_t) - j_{\bar{\alpha}_t}^\theta \right) \right] = n^{-1} E_x^{\pi^1, \pi_*^2} \left[\sum_{t=0}^{n-1} \left(h_{\bar{\alpha}_t}^\theta(x_t) - \bar{\alpha}_t h_{\bar{\alpha}_t}^\theta(x_{t+1}) \right) \right]$$

$$+ n^{-1} E_x^{\pi^1, \pi_*^2} \left[\sum_{t=0}^{n-1} \mathcal{L}_t \right].$$

Hence, from (1.13) and (4.13)

$$J(x, \pi^1, \pi_*^2) - j^* = \liminf_{n \to \infty} \left\{ n^{-1} E_x^{\pi^1, \pi_*^2} \left[\sum_{t=0}^{n-1} \left(h_{\bar{\alpha}_t}^\theta(x_t) - \bar{\alpha}_t h_{\bar{\alpha}_t}^\theta(x_{t+1}) \right) \right] \right.$$

$$\left. + n^{-1} E_x^{\pi^1, \pi_*^2} \left[\sum_{t=0}^{n-1} \mathcal{L}_t \right] \right\}. \tag{4.43}$$

Therefore, the remainder of the proof consists in showing that

$$\liminf_{n \to \infty} \left\{ n^{-1} E_x^{\pi^1, \pi_*^2} \left[\sum_{t=0}^{n-1} \left(h_{\bar{\alpha}_t}^\theta(x_t) - \bar{\alpha}_t h_{\bar{\alpha}_t}^\theta(x_{t+1}) \right) \right] + n^{-1} E_x^{\pi^1, \pi_*^2} \left[\sum_{t=0}^{n-1} \mathcal{L}_t \right] \right\} \le 0. \tag{4.44}$$

First observe that Condition $\bar{F}$.2 implies that $\{\bar{\alpha}_t\}$ remains constant for long time periods. Then, by applying similar arguments as in the proof of (3.40) and (3.41) in Sect. 3.3.1, we get

$$\lim_{n \to \infty} n^{-1} E_x^{\pi^1, \pi_*^2} \left[\sum_{t=0}^{n-1} \left(h_{\bar{\alpha}_t}^\theta(x_t) - \bar{\alpha}_t h_{\bar{\alpha}_t}^\theta(x_{t+1}) \right) \right] = 0. \tag{4.45}$$

Now, we will proceed to prove that

$$\lim_{n \to \infty} n^{-1} E_x^{\pi^1, \pi_*^2} \left[\sum_{t=0}^{n-1} \mathcal{L}_t \right] \le 0. \tag{4.46}$$

To this end, it is enough to prove that

$$\limsup_{t \to \infty} E_x^{\pi^1, \pi_*^2} [\mathcal{L}_t] \le 0.$$

Observe that, from (4.33), for each $t \in \mathbb{N}_0$,

$$\left| \int_{\Re^k} h^\theta_{\bar{\alpha}_t}(F(x,a,b,s))\theta_t(ds) - \int_{\Re^k} h^\theta_{\bar{\alpha}_t}(F(x,a,b,s))\theta(ds) \right| \leq \bar{\Delta}_t W(x).$$

Hence, adding and subtracting the terms

$$\bar{\alpha}_t \int_{\Re^k} h^\theta_{\bar{\alpha}_t}[F(x_t,\pi_t^1,\bar{\varphi}_t^2,s)]\theta_t(ds) \quad \text{and} \quad \bar{\alpha}_t \int_{\Re^k} h^{\theta_t}_{\bar{\alpha}_t}[F(x_t,\pi_t^1,\bar{\varphi}_t^2,s)]\theta_t(ds)$$

we get

$$\mathscr{L}_t \leq \bar{\Delta}_t W(x_t) + \mathscr{L}_t^0 + \mathscr{L}_t^1, \tag{4.47}$$

where

$$\mathscr{L}_t^0 := \left| \int_{\Re^k} h^\theta_{\bar{\alpha}_t}[F(x_t,\pi_t^1,\bar{\varphi}_t^2,s)]\theta_t(ds) - \int_{\Re^k} h^{\theta_t}_{\bar{\alpha}_t}[F(x_t,\pi_t^1,\bar{\varphi}_t^2,s)]\theta_t(ds) \right|,$$

$$\mathscr{L}_t^1 := r(x_t,\pi_t^1,\bar{\varphi}_t^2) + \bar{\alpha}_t \int_{\Re^k} h^{\theta_t}_{\bar{\alpha}_t}[F(x_t,\pi_t^1,\bar{\varphi}_t^2,s)]\theta_t(ds) - j^\theta_{\bar{\alpha}_t} - h^\theta_{\bar{\alpha}_t}(x_t).$$

Note that $\mathscr{L}_t^0 \leq \left\| h^{\theta_t}_{\bar{\alpha}_t} - h^\theta_{\bar{\alpha}_t} \right\|_W$ and, therefore, from (4.38),

$$\lim_{t\to\infty} E_x^{\pi^1,\pi_*^2} \mathscr{L}_t^0 = 0. \tag{4.48}$$

For $\mathscr{L}_t^1$, adding and subtracting $j^{\theta_t}_{\bar{\alpha}_t}$ and $h^{\theta_t}_{\bar{\alpha}_t}(x_t)$, from the definition of $\bar{\varphi}_t^2$ (see (4.34)), we obtain

$$\mathscr{L}_t^1 \leq \max_{\varphi^1 \in \mathbb{A}(x)} \left[r(x_t,\varphi^1,\bar{\varphi}_t^2) + \bar{\alpha}_t \int_S h^{\theta_t}_{\bar{\alpha}_t}[F(x_t,\varphi^1,\bar{\varphi}_t^2,s)]\theta_t(ds) \right] - j^{\theta_t}_{\bar{\alpha}_t} - h^{\theta_t}_{\bar{\alpha}_t}(x_t)$$

$$+ j^{\theta_t}_{\bar{\alpha}_t} - j^\theta_{\bar{\alpha}_t} + h^{\theta_t}_{\bar{\alpha}_t}(x_t) - h^\theta_{\bar{\alpha}_t}(x_t)$$

$$\leq \left| j^{\theta_t}_{\bar{\alpha}_t} - j^\theta_{\bar{\alpha}_t} \right| + \left\| h^{\theta_t}_{\bar{\alpha}_t} - h^\theta_{\bar{\alpha}_t} \right\|_W W(x_t).$$

Thus, (4.39) and (4.40) imply

$$\limsup_{t\to\infty} E_x^{\pi^1,\pi_*^2} \mathscr{L}_t^1 \leq 0. \tag{4.49}$$

Combining (4.41), (4.47), (4.48), and (4.49), we get (4.46), which, together with (4.45), yields (4.44). Thus, from (4.43)

$$J(x,\pi^1,\pi_*^2) \leq j^* \quad \forall x \in X.$$

Finally, since $\pi^1 \in \Pi^1$ is arbitrary, from Theorem 4.5,

$$j^* = \sup_{\pi^1 \in \Pi^1} J(x, \pi^1, \pi_*^2) \quad \forall x \in X.$$

The optimality of π_*^1 is proved similarly.

Finally, the average optimality of the pair $(\hat{\pi}_*^1, \hat{\pi}_*^2) \in \Pi^1 \times \Pi^2$ is proved following similar arguments as in part (c) of Theorem 4.8. ∎

4.5 Empirical Recursive Methods

According to Theorems 4.6 and 4.8, the empirical procedure to approximate the value V_α^θ and obtain an optimal pair of strategies is as follows. At each stage $t \in \mathbb{N}$, the players determine the distribution θ_t and select actions that would be optimal if θ_t were the true distribution. Hence, letting t enough large we obtain optimality in the original game, as is stated in Theorem 4.8. In order to perform in this way, the players must solve the equation $T_{(\alpha, \theta_t)} V_\alpha^{\theta_t} = V_\alpha^{\theta_t}$ for $V_\alpha^{\theta_t}$, and then determine the selectors φ_t^1 and φ_t^2 satisfying (4.17) and (4.18). However, solving such an equation could be difficult, which represents an obstacle to the implementation of the procedure. In order to avoid this inconvenience, we now present a value iteration algorithm which combined with the empirical process defines a recursive method to approximate the value of the game.

We define the sequence of functions $\{V_t\} \subset \mathbb{C}_W$ as $V_0 = 0$, and for $t \in \mathbb{N}$ and $x \in X$,

$$
\begin{aligned}
V_t(x) &= T_{(\alpha, \theta_t)} V_{t-1}(x) \\
&= \max_{\varphi^1 \in \mathbb{A}(x)} \min_{\varphi^2 \in \mathbb{B}(x)} \left[r(x, \varphi^1, \varphi^2) + \alpha \int_S V_{t-1}[F(x, \varphi^1, \varphi^2, s)] \theta_t(ds) \right] \\
&= \max_{\varphi^1 \in \mathbb{A}(x)} \min_{\varphi^2 \in \mathbb{B}(x)} \left[r(x, \varphi^1, \varphi^2) + \frac{\alpha}{t} \sum_{i=0}^{t-1} V_{t-1}[F(x, \varphi^1, \varphi^2, \xi_i)] \right]. \quad (4.50)
\end{aligned}
$$

Moreover, from Theorem 4.4 (see Theorem 4.6(b)), there exists $(\tilde{\varphi}_t^1, \tilde{\varphi}_t^2) \in \Phi^1 \times \Phi^2$ such that, for all $x \in X$,

$$
\begin{aligned}
V_t(x) &= r(x, \tilde{\varphi}_t^1, \tilde{\varphi}_t^2) + \frac{\alpha}{t} \sum_{i=0}^{t-1} V_{t-1}[F(x, \tilde{\varphi}_t^1, \tilde{\varphi}_t^2, \xi_i)] \\
&= \max_{\varphi^1 \in \mathbb{A}(x)} \left[r(x, \varphi^1, \tilde{\varphi}_t^2) + \frac{\alpha}{t} \sum_{i=0}^{t-1} V_{t-1}[F(x, \varphi^1, \tilde{\varphi}_t^2, \xi_i)] \right] \\
&= \min_{\varphi^2 \in \mathbb{B}(x)} \left[r(x, \tilde{\varphi}_t^1, \varphi^2) + \frac{\alpha}{t} \sum_{i=0}^{t-1} V_{t-1}[F(x, \tilde{\varphi}_t^1, \varphi^2, \xi_i)] \right].
\end{aligned}
$$

Similarly as Sect. 4.2.2, there exist $\tilde{\varphi}_\infty^1 \in \Phi^1$ and $\tilde{\varphi}_\infty^2 \in \Phi^2$ such that $\tilde{\varphi}_\infty^1(x, \omega) \in \mathbb{A}(x)$ and $\tilde{\varphi}_\infty^2(x, \omega) \in \mathbb{B}(x)$ are accumulation points of $\{\tilde{\varphi}_t^1(x, \omega)\}$ and $\{\tilde{\varphi}_t^2(x, \omega)\}$, respectively.

We define the strategies $\tilde{\pi}_*^i = \{\tilde{\varphi}_t^i\} \in \Pi^i$ and $\tilde{\pi}_\infty^i = \{\tilde{\varphi}_\infty^i\} \in \Pi_s^i$, for $i = 1, 2$.

The hope is that the pairs $(\tilde{\pi}_\infty^1, \tilde{\pi}_\infty^2)$ and $(\tilde{\pi}_*^1, \tilde{\pi}_*^2)$ have a good performance in the original game model $\mathscr{G}\mathscr{M}$, provided that V_t, for t enough large, gives a good approximation of the value function V_α^θ. These properties are established precisely in the following result.

Theorem 4.11. *Under Assumptions 4.1, 4.3, and 4.7*

(a) $\left\| V_t - V_\alpha^\theta \right\|_W \to 0 \;\; P-a.s., \; as \; t \to \infty;$

(b) the pair of strategies $(\tilde{\pi}_*^1, \tilde{\pi}_*^2) \in \Pi^1 \times \Pi^2$ is asymptotically optimal (see Definition 2.1);

(c) the pair of strategies $(\tilde{\pi}_\infty^1, \tilde{\pi}_\infty^2) \in \Pi_s^1 \times \Pi_s^2$ is optimal for the game $\mathscr{G}\mathscr{M}$.

Proof. Since for each $\alpha \in (0, 1)$ and $\mu \in \mathbb{P}(S)$, the operator $T_{(\alpha,\mu)}$ is contraction (see (4.6)), from Theorem 4.4 and (4.50) we have

$$\left\| V_\alpha^\theta - V_{t+1} \right\|_{\tilde{W}} \leq \left\| T_{(\alpha,\theta)} V_\alpha^\theta - T_{(\alpha,\theta_t)} V_\alpha^\theta \right\|_{\tilde{W}} + \left\| T_{(\alpha,\theta_t)} V_\alpha^\theta - T_{(\alpha,\theta_t)} V_t \right\|_{\tilde{W}}$$

$$\leq \left\| T_{(\alpha,\theta)} V_\alpha^\theta - T_{(\alpha,\theta_t)} V_\alpha^\theta \right\|_{\tilde{W}} + \gamma_\alpha \left\| V_\alpha^\theta - V_t \right\|_{\tilde{W}}$$

$$\leq \Delta_t + \gamma_\alpha \left\| V_\alpha^\theta - V_t \right\|_{\tilde{W}}, \tag{4.51}$$

where the last inequality follows from (4.22). Letting $l := \limsup_{t \to \infty} \left\| V_\alpha^\theta - V_t \right\|_{\tilde{W}} < \infty$ and taking $\limsup$ as t goes to infinity in (4.51), from (4.20) we obtain $0 \leq l \leq \gamma_\alpha l$. Thus, as $\gamma_\alpha \in (\alpha, 1)$, we have that $l = 0$, which, together with (4.7), yields part (a).

The asymptotic optimality of the pair $(\tilde{\pi}_*^1, \tilde{\pi}_*^2)$ as well as the optimality of $(\tilde{\pi}_\infty^1, \tilde{\pi}_\infty^2)$ are proved by following the ideas in the proofs of Theorem 2.11, adapted to the empirical distribution, and Theorem 4.8, respectively. ∎

The Average Case. Once we have the method to approximate the discounted value function V_α^θ, we can obtain a procedure to approximate the value of the average game and optimal strategies by applying the VDFA introduced in Sect. 3.2.

Observe that for each $\alpha \in (0, 1)$, $x \in X$, and $\theta \in \mathbb{P}(S)$, the equations

$$V_\alpha^\theta(x) = T_{(\alpha,\theta)} V_\alpha^\theta(x)$$

and

$$j_\alpha^\theta + h_\alpha^\theta(x) = T_{(\alpha,\theta)} h_\alpha^\theta(x)$$

are equivalent, where for a fixed state $z \in X$ (see (4.11))

$$j_\alpha^\theta := (1-\alpha)V_\alpha^\theta(z), \quad h_\alpha^\theta(x) := V_\alpha^\theta(x) - V_\alpha^\theta(z).$$

Then, the strategies $\pi_*^i = \{\bar{\varphi}_t^i\}$ in Theorem 4.10 are defined by the selectors $\bar{\varphi}_t^i$, for $i = 1, 2$, such that

$$V_{\bar{\alpha}_t}^{\theta_t}(x) = T_{(\bar{\alpha}_t,\theta_t)} V_{\bar{\alpha}_t}^{\theta_t}(x)$$

$$= r(x, \bar{\varphi}_t^1, \bar{\varphi}_t^2) + \bar{\alpha}_t \int_S V_{\bar{\alpha}_t}^{\theta_t}[F(x, \bar{\varphi}_t^1, \bar{\varphi}_t^2, s)] \theta_t(ds)$$

$$= \max_{\varphi^1 \in \mathbb{A}(x)} \left[r(x, \varphi^1, \bar{\varphi}_t^2) + \bar{\alpha}_t \int_S V_{\bar{\alpha}_t}^{\theta_t}[F(x, \varphi^1, \bar{\varphi}_t^2, s)] \theta_t(ds) \right]$$

$$= \min_{\varphi^2 \in \mathbb{B}(x)} \left[r(x, \bar{\varphi}_t^1, \varphi^2) + \bar{\alpha}_t \int_S V_{\bar{\alpha}_t}^{\theta_t}[F(x, \bar{\varphi}_t^1, \varphi^2, s)] \theta_t(ds) \right],$$

where $\{\bar{\alpha}_t\}$ is a sequence of discount factors satisfying Conditions $\bar{F}.1$ and $\bar{F}.2$, and $V_{\bar{\alpha}_t}^{\theta_t}$ is the value of the game $\mathscr{GM}_t^{\bar{\alpha}_t}$. According to these facts and Theorem 4.10, in order to approximate the value of the game and optimal strategies for the average criterion, a first step is to approximate the function $V_{\bar{\alpha}_t}^{\theta_t}$. To this end, we can apply, for each $t \in \mathbb{N}_0$, the empirical value iteration procedure stated in Theorem 4.11. Let $\{V_n^{(t)}\} \subset \mathbb{C}_W$ be the sequence of functions defined as

$$V_0^{(t)} = 0; \tag{4.52}$$

$$V_n^{(t)}(x) = T_{(\bar{\alpha}_t,\theta_n)} V_{n-1}^{(t)}(x), \quad n \in \mathbb{N}, x \in X.$$

It is clear that from Theorem 4.11(a), for each $t \in \mathbb{N}_0$,

$$\left\| V_n^{(t)} - V_{\bar{\alpha}_t}^\theta \right\|_W \to 0, \quad \text{as } n \to \infty. \tag{4.53}$$

In addition, observe that

$$\left\| V_n^{(t)} - V_{\bar{\alpha}_t}^{\theta_t} \right\|_W \le \left\| V_n^{(t)} - V_{\bar{\alpha}_t}^\theta \right\|_W + \left\| V_{\bar{\alpha}_t}^\theta - V_{\bar{\alpha}_t}^{\theta_t} \right\|_W.$$

Hence, combining this fact with (4.53) and (4.37), we get, for all $(\pi^1, \pi^2) \in \Pi^1 \times \Pi^2$,

$$\lim_{t \to \infty} \lim_{n \to \infty} E_x^{\pi^1,\pi^2} \left\| V_n^{(t)} - V_{\bar{\alpha}_t}^{\theta_t} \right\|_W = 0. \tag{4.54}$$

Similarly, observe that

$$E_x^{\pi^1,\pi^2}\left|j_{\bar{\alpha}_t}^{\theta_t}-j^*\right| \leq E_x^{\pi^1,\pi^2}\left|j_{\bar{\alpha}_t}^{\theta_t}-j_{\bar{\alpha}_t}^{\theta}\right|+\left|j_{\bar{\alpha}_t}^{\theta}-j^*\right|.$$

Thus, from (4.39) and (4.13), we have

$$\lim_{t\to\infty} E_x^{\pi^1,\pi^2}\left|j_{\bar{\alpha}_t}^{\theta_t}-j^*\right|=0,$$

that is

$$\lim_{t\to\infty} E_x^{\pi^1,\pi^2}\left|(1-\bar{\alpha}_t)V_{\bar{\alpha}_t}^{\theta_t}(z)-j^*\right|=0. \tag{4.55}$$

Therefore, from (4.54)

$$\lim_{t\to\infty}\lim_{n\to\infty} E_x^{\pi^1,\pi^2}\left|(1-\bar{\alpha}_t)V_n^{(t)}(z)-j^*\right|=0. \tag{4.56}$$

The use of empirical distribution defines a very general approximation method of the value function and construction of optimal strategies, since both the random variables ξ_t and the distribution θ can be arbitrary. This generality entails imposing equicontinuity and equi-Lipschitz conditions for the discounted and the average criteria, respectively. In fact, even in the case of bounded payoff function r, such conditions are necessary because we need uniform convergence on $(x,a,b) \in \mathbb{K}$ (see Propositions 4.1 and 4.2).

It is worth remarking that to obtain estimation and control procedures under the empirical distribution it is not necessary to implement a projection scheme as the case of density estimation studied in previous chapters (see (2.38) and (3.17)).

Chapter 5
Difference-Equation Games: Examples

In this chapter we introduce several examples to show the relevance of the estimation and control procedures analyzed throughout the book.

In the first part we will focus on illustrating the assumptions on the game models, which can be classified into five classes, specifically:

- *continuity conditions* contained in Assumptions 2.1, 2.2, and 2.3;
- *W-growth conditions* in Assumptions 2.4 and 2.7, and their variants (see Assumptions 2.8, 3.1, 3.4, 4.1, and 4.3);
- *conditions on the density* stated in Assumptions 2.9 and 2.10;
- *ergodicity conditions* given in Assumption 3.2 (see also Assumption 4.2) to analyze the average criterion;

  and finally

- *equicontinuity and equi-Lipschitz conditions* in Assumptions 4.7 and 4.9 used to formulate the empirical estimation-approximation schemes.

All these assumptions will be illustrated in zero-sum games evolving according to a stochastic difference equation of the form

$$x_{t+1} = F(x_t, a_t, b_t, \xi_t) \text{ for } t \in \mathbb{N}_0, \tag{5.1}$$

where $F : \mathbb{K} \times S \to X$ is a given measurable function, assuming compactness of the admissible actions sets $A(x)$ and $B(x)$, for $x \in X$. In addition, $\{\xi_t\}$ is a sequence of observable i.i.d. random variables defined on a probability space $(\Omega, \mathscr{F}, P)$, taking values in a Borel space S, with common distribution $\theta \in \mathbb{P}(S)$, and independent of the initial state x_0.

J. A. Minjárez-Sosa, *Zero-Sum Discrete-Time Markov Games with Unknown Disturbance Distribution*, SpringerBriefs in Probability and Mathematical Statistics, https://doi.org/10.1007/978-3-030-35720-7_5

We conclude presenting numerical implementations of the estimation and control algorithms in two examples. The first one is a class of linear-quadratic games where a suitable density estimation process is applied, and the optimal strategies are found in the set of pure strategies (see Definition 1.2). Next, the empirical approximation is showed in a single example with finite state space.

5.1 Continuity Conditions

In this section we present some insights on the fulfillment of the continuity condition of the integral

$$(x,a,b) \rightarrow \int_X v(y)Q(dy|x,a,b) = \int_S v[F(x,a,b,s)]\theta(ds) \qquad (5.2)$$

for $(x,a,b) \in \mathbb{K}$, for the cases $v \in \mathbb{B}(X)$ and $v \in \mathbb{C}(X)$, involved in Assumptions 2.1, 2.2, and 2.3. Essentially, the continuity in (5.2) depends on the properties of the function F in (5.1) and of the distribution $\theta \in \mathbb{P}(S)$ (see Propositions C.3 and C.4 in Appendix C).

Example 5.1. Suppose that for every $s \in S$, the function $F(x,a,b,s)$ is continuous in $(x,a,b) \in \mathbb{K}$, and $\{\xi_t\}$ is a sequence of i.i.d. r.v. independent on $x_0 \in X$, with distribution $\theta \in \mathbb{P}(S)$. Then, for every $v \in \mathbb{C}(X)$, the function $v[F(\cdot,\cdot,\cdot,s)]$ is continuous, which implies, by the Dominated Convergence Theorem, that

$$(x,a,b) \rightarrow \int_S v[F(x,a,b,s)]\theta(ds), \quad (x,a,b) \in \mathbb{K}, \qquad (5.3)$$

is continuous. ∎

Example 5.2 (Additive Noise Systems). Consider stochastic games evolving on $X = \mathfrak{R}$ according to an additive noise difference equation of the form

$$x_{t+1} = F(x_t,a_t,b_t,\xi_t) = G(x_t,a_t,b_t) + \xi_t \text{ for } t \in \mathbb{N}_0,$$

with $A = B = \mathfrak{R}$, where G is a continuous function and $\{\xi_t\}$ is a sequence of i.i.d. r.v. with a continuous density ρ on $S = \mathfrak{R}$ with respect to Lebesgue measure. In this setting, for every $v \in \mathbb{B}(X)$, applying Scheffé's Theorem (see Theorem D.1), the function

$$(x,a,b) \rightarrow \int_X v(y)Q(dy|x,a,b) = \int_\mathfrak{R} v[G(x,a,b)+s]\rho(s)ds$$

$$= \int_\mathfrak{R} v(y)\rho[G(x,a,b)-s]ds, \quad (x,a,b) \in \mathbb{K}, \qquad (5.4)$$

is continuous. ∎

In the following sections we present examples of zero-sum games whose dynamics fall within one of the cases (5.3) or (5.4) or both.

5.2 Autoregressive Game Models

Let $\{x_t\}$ be the state process of a game whose dynamic is defined by the equation

$$x_{t+1} = G(a_t, b_t)x_t + \xi_t \text{ for } t \in \mathbb{N}_0,$$

with initial state x_0, state space $X = [0, \infty)$, compact actions sets $A(x) = A \subset \mathfrak{R}$, and $B(x) = B \subset \mathfrak{R}$ for $x \in X$, and $G : A \times B \to (0, \lambda]$ is a given continuous function with $\lambda < 1$. Moreover, $\{\xi_t\}$ is a sequence of i.i.d. random variables taking values in $S = [0, \infty)$ with a continuous density ρ and finite expectation $E[\xi_0] = \bar{\xi}$.

In the spirit of a zero-sum game model, we can think in the following situation. Let $x^* \in X$ be a fixed state. Suppose that the objective of player 1 is to select actions tending to move the game process $\{x_t\}$ away from x^*, while player 2 wants to move the process $\{x_t\}$ as close as possible to x^*. In these circumstances, we consider a payoff function of the form

$$r(x, a, b) = \sqrt{|x - x^*|}, \quad (x, a, b) \in \mathbb{K}.$$

In addition, defining $W(x) := \sqrt{x + x^* + 1}, x \in X$, we get

$$\int_X W^2(y)Q(dy|x, a, b) = \int_0^\infty W^2[G(a, b)x + s]\rho(s)ds$$

$$= \int_0^\infty [G(a, b)x + s + x^* + 1]\rho(s)ds$$

$$\leq \lambda x + x^* + 1 + \bar{\xi}$$

$$\leq \lambda(x + x^* + 1) + x^* + \bar{\xi} + 1$$

$$= \lambda_0 W^2(x) + d_0,$$

where $\lambda_0 := \lambda$ and $d_0 := x^* + \bar{\xi} + 1$. Moreover, letting $\beta := \lambda_0^{1/2}$ and $d := d_0^{1/2}$ we get (see (2.23) in Remark 2.4 (a))

$$\int_0^\infty W[G(a, b)x + s]\rho(s)ds \leq \beta W(x) + d.$$

On the other hand, applying similar arguments to (5.4) we can prove that the function

$$(x,a,b) \rightarrow \int_0^\infty W[G(a,b)x+s]\rho(s)ds, \quad (x,a,b) \in \mathbb{K},$$

is continuous. Hence, Assumptions 2.4 and 2.8 with $p = 2$ are satisfied (see Assumptions 3.1 (d), (e), 3.4, and 4.1).

To verify Assumptions 2.9 and 2.10, observe that

$$\psi(s) := \sup_{x \in X} \sup_{a \in A(x)} \sup_{b \in B(x)} \frac{1}{W(x)} W[F(x,a,b,s)]$$

$$= \sup_{x \in X} \sup_{a \in A(x)} \sup_{b \in B(x)} \left(\frac{G(a,b)x+s+x^*+1}{x+x^*+1} \right)^{1/2} < (1+s)^{1/2}, \quad s \in [0,\infty).$$

Thus, taking $\widetilde{\rho}(\cdot) \equiv \rho(\cdot)$ we obtain

$$\bar{\Psi} := \int_0^\infty \psi^2(s)\widetilde{\rho}(s)ds \leq \int_0^\infty (1+s)\rho(s)ds = 1 + \bar{\xi} < \infty.$$

In the particular case in which $\{\xi_t\}$ is assumed to be a sequence of i.i.d. random variables taking values in $S = [0,s^*]$, for some $s^* > 0$, Assumption 4.3 is satisfied. Indeed, for all $(x,a,b) \in \mathbb{K} \times S$,

$$W^2[G(a,b)x+s] = G(a,b)x+s+x^*+1$$

$$\leq \lambda x + s^* + x^* + 1$$

$$\leq \lambda(x+x^*+1) + s^* + x^* + 1.$$

Hence
$$W[G(a,b)x+s] \leq \lambda^{1/2}W(x) + (s^*+x^*+1)^{1/2}.$$

Therefore, defining $\beta := \lambda^{1/2}$ and $d := (s^*+x^*+1)^{1/2}$, Assumption 4.3 holds.

In conclusion, from Theorem 2.11, there exists a pair $(\pi_*^1, \pi_*^2) \in \Pi^1 \times \Pi^2$ of asymptotically discounted optimal strategies.

5.3 Linear-Quadratic Games

The linear-quadratic games, known as LQ-games, are dynamic games evolving according to a linear equation with quadratic payoff (see [14]). We consider the following particular case. Let $\{x_t\}$ be the state process satisfying the equation

$$x_{t+1} = x_t + a_t + b_t + \xi_t \quad \text{for } t \in \mathbb{N}_0,$$

with x_0 given, where $X = A = B = \Re$. The random disturbance process $\{\xi_t\}$ is a sequence of i.i.d. random variables normally distributed with mean zero and unknown variance $\sigma^2 \in (0,3)$, that is

$$\rho(s) := \frac{1}{\sigma\sqrt{2\pi}} \exp\left(-\frac{s^2}{2\sigma^2}\right) \quad \forall s \in \Re,\ E(\xi_t) = 0,\ \text{and}\ \sigma^2 = E(\xi_t^2),\ \forall t \in \mathbb{N}_0.$$

$$(5.5)$$

The sets of admissible actions for players 1 and 2 are $A(x) = B(x) = [-|x|/2, |x|/2]$, and the one-stage payoff r is the quadratic function

$$r(x,a,b) = x^2 + a^2 - b^2.$$

Observe that for $a,b \in [-|x|/2, |x|/2]$ we have

$$r(x,a,b) \le x^2 + a^2 + b^2 \le x^2 + \frac{1}{2}x^2$$

$$\le \frac{3}{2}(x^2 + 1) = M(x^2 + 1),$$

with $M = \dfrac{3}{2}$.

Defining $W(x) := x^2 + 1$, for all $(x,a,b) \in \mathbb{K}$, we have

$$\int_\Re W(y) Q(dy|x,a,b) = \int_\Re [(x+a+b+s)^2 + 1] \frac{1}{\sigma\sqrt{2\pi}} \exp\left(-\frac{s^2}{2\sigma^2}\right) ds$$

$$= \int_\Re (x+a+b+s)^2 \frac{1}{\sigma\sqrt{2\pi}} \exp\left(-\frac{s^2}{2\sigma^2}\right) ds + 1$$

$$= \int_\Re y^2 \frac{1}{\sigma\sqrt{2\pi}} \exp\left(-\frac{(y-(x+a+b))^2}{2\sigma^2}\right) dy + 1$$

$$= (x+a+b)^2 + \sigma^2 + 1 \le 4x^2 + 4 \le 4W(x). \qquad (5.6)$$

Hence, Assumption 2.7 holds with $\tilde{\gamma} = 4$ and any discount factor $\alpha < \frac{1}{4}$.

On the other hand, observe that the function ψ in Assumption 2.10 satisfies that

$$\psi(s) = \sup_{x\in X} \sup_{a\in A(x)} \sup_{b\in B(x)} \frac{(x+a+b+s)^2 + 1}{x^2 + 1}$$

$$= \sup_{x\in X} \frac{(2x+s)^2 + 1}{x^2 + 1}$$

$$\le s^2 + \sup_{x\in X} \frac{4xs}{x^2 + 1} + 5$$

$$< s^2 + 2s + 5,$$

after observing that

$$\sup_{x \in X} \frac{4x}{x^2+1} = 2.$$

Furthermore, assuming that $\widetilde{\rho}$ is of the form $\widetilde{\rho}(s) \equiv k_1 \exp(-k_2 |s|)$, $s \in \Re$, we can choose the positive constants k_1 and k_2 such that

$$\rho(s) \le \widetilde{\rho}(s) \ \forall s \in \Re, \tag{5.7}$$

and

$$\bar{\Psi} := \int_{\Re} \psi^2(s)\widetilde{\rho}(s)ds \le \int_{\Re} (s^2 + 2s + 5)k_1 \exp(-k_2 |s|)ds < \infty.$$

That is, Assumptions 2.9 and 2.10 hold.

Considering that Assumption 2.4(b) can be replaced by the assumption 2.7 (see (2.11)), Theorem 2.11 yields the existence of an asymptotically discounted optimal pair of strategies.

Similarly as (5.6), if we consider the function $W(x) := \left(x^2+1\right)^{1/2}$, for all $(x,a,b) \in \mathbb{K}$, we can prove that Assumption 2.8 is satisfied with $p = 2$.

5.4 A Game Model for Reservoir Operations

In this section we analyze a model of a single reservoir with infinite capacity described as follows.

There are two purposes in the operation of the reservoir system: the *social* related to the water provision to meet the demand for domestic use, and the *economic* where water is used in hydropower generation, land irrigation, etc. (see, e.g., [77, 78]). Under certain circumstances (e.g., political, economic, and/or social), satisfying these requirements leads to conflict situations in such a way that these can be assumed as opposing objectives, in the sense that water used for one purpose is considered as water loss for the other.

Let $\xi_{1,t}$ be a random variable representing the inflows into the reservoir which happen at nonnegative random times $T_t, t \in \mathbb{N}$. At each time T_t, the decision maker observes the reservoir level $x_t = x \in X := [0, \infty)$ and selects actions $a_t = a \in A := [0, a^*]$ and $b_t = b \in B := [b_*, b^*]$, representing the consumption rates to satisfy both the water demand for social purposes and for economic purposes, respectively, where a_*, b_*, and b^* are fixed positive constants.

Observe that the reservoir level at time T_{t+1} depends on the inflow $\xi_{1,t+1}$ and the volume of water consumed W_t during period $[T_t, T_{t+1})$. For instance, if the volume of water $x_t = x$ is not depleted before time T_{t+1}, then

$$W_t = (a+b)\xi_{2,t+1}, \tag{5.8}$$

where $\xi_{2,t+1} := T_{t+1} - T_t, t \in \mathbb{N}_0$, and $T_0 := 0$. Hence, the reservoir level at time T_{t+1} is

$$x_{t+1} = x - W_t + \xi_{1,t+1}. \tag{5.9}$$

On the other hand, if the reservoir level $x_t = x$ is depleted before time T_{t+1}, then

$$x_{t+1} = \xi_{1,t+1}. \tag{5.10}$$

Combining (5.8)–(5.10), the reservoir level process $\{x_t\}$ evolves according to the stochastic difference equation

$$x_{t+1} = \max(x_t - (a_t + b_t)\xi_{2,t+1}, 0) + \xi_{1,t+1}, \quad t \in \mathbb{N}_0,$$

where $x_0 = x \in X$ is the initial water volume. Hence, the dynamic is given by a function $F : \mathbb{K} \times \mathbb{R}_+^2 \to X$ defined as

$$F(x, a, b, s) := \max(x - (a+b)s_2, 0) + s_1,$$

for all $(x, a, b) \in \mathbb{K}$, $s := (s_1, s_2) \in \mathbb{R}_+^2 := \mathbb{R}^+ \times \mathbb{R}^+$, which is continuous.

In order to verify our assumptions, we assume that the random processes $\{\xi_{1,t}\}$ and $\{\xi_{2,t}\}$ satisfy the following conditions:

C1 The processes $\{\xi_{1,t}\}$ and $\{\xi_{2,t}\}$ are independent sequences of i.i.d. random variables. We denote $\rho_i(\cdot)$ the density of $\xi_{i,t}$ for $i \in \{1,2\}$.

C2 The joint density $\rho^*(\cdot, \cdot) = \rho_1(\cdot)\rho_2(\cdot)$ is continuous and bounded by the function

$$\tilde{\rho}(s_1, s_2) := \begin{cases} \tilde{L}\exp(l(s_1 + s_2)) \ if \ (s_1, s_2) \in S_0, \\ \\ 0 \qquad\qquad\qquad otherwise, \end{cases}$$

where $\tilde{L}$ and l are positive constants, and S_0 is a compact subset of $\mathbb{R}_+^2$.

C3 The processes $\{\xi_{1,t}\}$ and $\{\xi_{2,t}\}$ have finite expectation. Moreover

$$E\xi_{1,t} < b_* E\xi_{2,t}.$$

Let Θ_Y be the moment generating function of the random variable $Y := \xi_{1,t} - b_* E\xi_{2,t}$, that is

$$\Theta_Y(t) := E\exp(t(\xi_{1,t} - b_* E\xi_{2,t})) = \iint_{\mathbb{R}_+^2} \exp(t(s_1 - b_* s_2))\rho^*(s_1, s_2)ds_1 ds_2.$$

From Condition C3, observe that the derivative $\Theta_Y'(0) = E[Y] < 0$. Now, since $\Theta_Y(0) = 1$, there exists $t_0 > 0$ such that

$$\beta := \Theta_Y(t_0/2) < 1 \ \text{ and } \ \lambda_0 := \Theta_Y(t_0) < 1. \tag{5.11}$$

We proceed to verify Assumptions 3.1 and 3.2 (see Assumptions 4.1 and 4.2). To this end, we define

$$W(x) := L' \exp(t_0 x/2), \quad x \in X, \text{ for some } L' > 0;$$

$$\lambda(x,a,b) := P[\xi_{2,t} > x/(a+b)], \quad (x,a,b) \in \mathbb{K};$$

$$m^*(D) := P[\xi_{1,t} \in D], \quad D \in \mathscr{B}(X).$$

Let $r : \mathbb{K} \to \mathfrak{R}$ be a continuous one-stage payoff function such that

$$0 \leq r(x,a,b) \leq W(x), \quad \forall (x,a,b) \in \mathbb{K}.$$

In addition, from the continuity of the functions W and F, applying similar arguments as (5.3) we can prove that the function

$$(x,a,b) \to \iint_{\mathfrak{R}^2_+} W[F(x,a,b,s)] \rho^*(s_1,s_2) ds_1 ds_2, \quad (x,a,b) \in \mathbb{K}, \tag{5.12}$$

is continuous. Hence, Assumption 3.1 is satisfied.

For the Assumption 3.2, let $\bar{S} := \{(s_1,s_2) \in \mathfrak{R}^2_+ : x - (a+b)s_2 \leq 0\}$ and $\bar{S}^c := \mathfrak{R}^2_+ - A_1$. Then Conditions C1–C3 yield

$$\iint_{\bar{S}^c} W[F(x,a,b,s)] \rho^*(s_1,s_2) ds_1 ds_2 = \iint_{\bar{S}^c} W(x + s_1 - (a+b)s_2) \rho^*(s_1,s_2) ds_1 ds_2$$

$$\leq \iint_{\bar{S}^c} L' \exp\left(\frac{t_0}{2}(s_1 - (a+b)s_2)\right) \rho^*(s_1,s_2) ds_1 ds_2$$

$$\leq W(x) \iint_{\bar{S}^c} \exp\left(\frac{t_0}{2}(s_1 - b_* s_2)\right) \rho^*(s_1,s_2) ds_1 ds_2 \leq \beta W(x), \tag{5.13}$$

and

$$\iint_{\bar{S}} W[F(x,a,b,s)] \rho^*(s_1,s_2) ds_1 ds_2 = \iint_{\bar{S}} W(s_1) \rho^*(s_1,s_2) ds_1 ds_2$$

$$= \lambda(x,a,b) E[W(\xi_{1,t})]$$

$$= \lambda(x,a,b) \int_X W(y) m^*(dy). \tag{5.14}$$

Combining (5.13) and (5.14) we get

$$\int_X W(y)Q(dy|x,a,b) = \iint_{\mathfrak{R}^2_+} W[F(x,a,b,s)]\rho^*(s_1,s_2)ds_1ds_2$$

$$\leq \beta W(x) + \lambda(x,a,b)d, \tag{5.15}$$

where $d = \int_X W(y)m^*(dy) < \infty$.

On the other hand, note that

$$Q(D|x,a,b) = \iint_{\mathfrak{R}^2_+} 1_D(F(x,a,b,s_1,s_2))\rho^*(s_1,s_2)ds_1ds_2$$

$$= \iint_{\bar{S}} 1_D(s_1)\rho^*(s_1,s_2)ds_1ds_2$$

$$+ \iint_{\bar{S}^c} 1_D(x+s_1-(a+b)s_2)\rho^*(s_1,s_2)ds_1ds_2$$

$$\geq \iint_{\bar{S}} 1_D(s_1)\rho^*(s_1,s_2)ds_1ds_2$$

$$= \lambda(x,a,b)m^*(D), \quad D \in \mathscr{B}(X).$$

Moreover,

$$\Pr[b_*\xi_{2,t} > \xi_{1,t}] = \int_0^\infty P[b_*\xi_{2,t} > y]\rho_1(y)dy$$

$$= \int_0^\infty P[b_*\xi_{2,t} > x]m^*(dx),$$

which is positive due to Condition C3. Since

$$Q(D|x,a,b) \geq P[\xi_{2,t} > x/(a+b)]m^*(D)$$
$$\geq P[\xi_{2,t} > x/b_*]m^*(D),$$

we see that the ergodicity Assumption 3.2 holds.

Observe that the inequality (3.14) in Assumption 3.4 follows by using similar arguments as in (5.15) with $p = 2$, $\lambda_0 := \Theta_Y(t_0)$ (see (5.11)) and $d_0 := E[W^2(\xi_{1,t})]$, which is finite because of the continuity of W and Condition C2.

Finally, observe that

$$\Psi(s) \leq \max\{\exp(t_0 s_1/2), \exp(t_0/2(s_1 - b_* s_2))\}, \quad s = (s_1, s_2).$$

Hence, from Condition C2,

$$\iint_{\Re_+^2} \Psi^2(s_1, s_2)\widetilde{\rho}(s_1, s_2)ds_1 ds_2 = \iint_{S_0} \Psi^2(s_1, s_2)\left(\widetilde{\rho}(s_1, s_2)\right)ds_1 ds_2 < \infty,$$

which implies Assumption 3.4.

Therefore we can conclude, from Theorem 3.5, that there exists an average optimal pair of strategies.

5.5 A Storage Game Model

We now introduce a zero-sum game model to analyze a class of storage systems with controlled input/output, whose evolution in time is as follows.

At time T_t when an amount of a certain product $R > 0$ accumulates for admission in the system, player 1 selects an action $a \in [a_*, 1] =: A$, with $a_* \in (0,1)$, representing the portion of R to be admitted. That is, aR is the product amount which player 1 has decided to admit to the system. There is a continuous consumption of the admitted product, controlled by player 2 when selecting $b \in [b_*, b^*] =: B$ ($0 < b_* < b^*$) which represents the consumption rate per unit time. Thus, if $x_t \in X := [0, \infty)$ is the stock level, a_t and b_t are the decisions of players 1 and 2, respectively, at the time of the t-th decision epoch T_t, then the process $\{x_t\}$ evolves according to the equation

$$x_{t+1} = (x_t + a_t R - b_t \xi_{t+1})^+,$$

where $\xi_{t+1} := T_{t+1} - T_t$, $t \in \mathbb{N}_0$. We suppose that $\{\xi_t\}$ is a sequence of i.i.d. random variables with unknown density ρ such that

$$E[\xi] > \frac{R}{b_*}. \tag{5.16}$$

We proceed as in Example 5.4 to verify Assumptions 3.1 and 3.2. Indeed, let Θ be the moment generating function of the random variable $R - b_* \xi$, that is,

$$\Theta(t) = E[\exp(t(R - b_* \xi))].$$

Observe that (5.16) implies that $\Theta'(0) < 1$. In addition, because $\Theta(0) = 1$, there exists $t_0 > 0$ such that

$$\beta := \Theta(t_0) = E[\exp(t_0(R - b_* \xi))] < 1. \tag{5.17}$$

We suppose that the one-stage payoff function r is an arbitrary continuous function on $\mathbb{K}$ such that

$$0 \leq r(x,a,b) \leq M e^{t_0 x}.$$

Hence, defining $W(x) := \exp(t_0 x)$, it is easy to prove that (see (5.12))

$$(x,a,b) \rightarrow \int_0^\infty W\left[(x+aR-bs)^+\right]\rho(s)ds, \quad (x,a,b) \in \mathbb{K},$$

is continuous, which yields Assumption 3.1.

Let

$$\lambda(x,a,b) := P[x+aR-b\xi \leq 0] \quad \text{for } (x,a,b) \in \mathbb{K},$$

and $m^*(\cdot) := \delta_0(\cdot)$, where δ_0 is the Dirac measure concentrated at $x=0$. Then, for $D \in \mathcal{B}(X)$,

$$Q(D|x,a,b) = \int_{\Re} 1_D(F(x,a,b,s))\rho(s)ds = \int_{\Re} 1_D((x+aR-bs)^+)\rho(s)ds$$

$$= P\left[(x+aR-b\xi)^+ \in D\right] \geq P[x+aR-b\xi \leq 0]\,\delta_0(D)$$

$$= \lambda(x,a,b)m^*(D). \tag{5.18}$$

In addition, for each $(x,a,b) \in \mathbb{K}$,

$$\int_X W(y)Q(dy \mid x,a,b) = \int_0^\infty \exp\left(t_0(x+aR-bs)^+\right)\rho(s)ds$$

$$= P[x+aR-b\xi \leq 0] + \exp(t_0 x)\int_0^\infty \exp(t_0(aR-bs))\rho(s)ds$$

$$\leq P[x+aR-b\xi \leq 0] + \exp(t_0 x)\int_0^\infty \exp(t_0(R-b_*s))\rho(s)ds$$

$$= \beta W(x) + \lambda(x,a,b). \tag{5.19}$$

On the other hand, observe that

$$\bar{\Lambda}(x) := \inf_{a \in A} \inf_{b \in B} \psi(x,a,b) = \inf_{a \in A} \inf_{b \in B} P[x+aR-b\xi \leq 0]$$

$$\geq P[x+R-b_*\xi \leq 0].$$

Thus,

$$\int_X \bar{\Lambda}(x)m^*(dx) \geq \int_X P[x+R-b_*\xi \leq 0]m^*(dx)$$

$$= P[R-b_*\xi \leq 0] > 0. \tag{5.20}$$

where the last inequality holds because (see (5.16))

$$E[R - b_*\xi] < 0.$$

Therefore (5.18)–(5.20) imply Assumption 3.2 (see Assumption 4.2).

In addition, observe that by the continuity of Θ, we can choose $p > 1$ such that (see (5.17))

$$\lambda_0 := \Theta(pt_0) = E[\exp(pt_0(R - b_*\xi))] < 1.$$

Thus, a straightforward calculation shows that (see (5.19))

$$\int_X W^p(y)Q(dy \mid x,a,b) \le P[x + aR - b\xi \le 0]$$
$$+ \exp(t_0 px) \int_0^\infty \exp(t_0 p(R - b_* s))\rho(s)ds,$$

which in turn implies

$$\int_X W^p(y)Q(dy \mid x,a,b) \le \lambda_0 W^p(x) + d_0, \qquad (5.21)$$

with $d_0 = 1$. Hence, Assumption 3.4 (a) is satisfied.

Now, for $s \in [0,\infty)$, let

$$K_1 := \{(x,a,b) \in \mathbb{K} : x + aR - bs \le 0\}$$

$$K_2 := \{(x,a,b) \in \mathbb{K} : x + aR - bs > 0\}.$$

Then

$$\psi(s) := \sup_{x\in X}\sup_{a\in A}\sup_{b\in B} \frac{1}{W(x)}W[F(x,a,b,s)]$$

$$= \sup_{x\in X}\sup_{a\in A}\sup_{b\in B} \frac{\exp(t_0(x + aR - bs)^+)}{\exp(t_0 x)}$$

$$= \max\left\{ \sup_{K_1} \frac{\exp(t_0(x + aR - bs)^+)}{\exp(t_0 x)}, \sup_{K_2} \frac{\exp(t_0(x + aR - bs)^+)}{\exp(t_0 x)} \right\}$$

$$= \max\left\{ \sup_{K_1}\exp(-t_0 x), \sup_{K_2}\exp(t_0(aR - bs)) \right\}$$

$$= \max\{1, \exp(t_0(aR - b_* s))\} < \infty, \quad s \in [0,\infty). \qquad (5.22)$$

In this sense, in order to verify Assumption 3.4(b), we assume that the density ρ is a function satisfying $\rho(s) \leq \widetilde{\rho}(s) := \widetilde{L}\exp(-qs)$, where $\widetilde{L}$ and q are suitable constants such that (2.29) holds. Hence, Assumption 3.4 is satisfied (see Assumption 2.10).

In conclusion, from Theorem 3.5, there exist average optimal strategies for players.

5.6 Equicontinuity and Equi-Lipschitz Conditions

The convergence properties of the empirical approximation-estimation algorithms introduced in Chap. 4 are strongly based on the equicontinuity and equi-Lipschitz conditions stated in Assumptions 4.7 and 4.9 for the discounted and average criteria, respectively. Such conditions are satisfied in several contexts. For instance, if the disturbance space S is countable, i.e., if the disturbance process $\{\xi_t\}$ is formed by discrete random variables, the equicontinuity is trivially satisfied with respect to the discrete topology. Now, considering that equi-Lipschitz implies equicontinuity, another set of less obvious conditions can be obtained applying the ideas from [16] or [23, 24] for MDPs, and by imposing Lipschitz-like conditions on the payoff function r and on the transition kernel Q.

Although we can obtain equicontinuity without going through the equi-Lipschitz property, in this section we introduce sufficient conditions for Assumption 4.9, which in turn implies Assumption 4.7 in the case $S = \mathfrak{R}^k$. We will do this under Assumptions 4.1 and 4.2, and the following conditions.

Assumption 5.1 *(a) For all $x \in X$, $A(x) = A$ and $B(x) = B$ are compact sets.*

(b) The one-stage payoff function r and the function F are Lipschitz functions in the following sense: For all $x, x' \in X$ and a metric d_X on X,

$$\sup_{(a,b)\in A\times B} \left| r(x,a,b) - r(x',a,b) \right| \leq L_r d_X(x,x')$$

and

$$\sup_{(a,b)\in A\times B} d_X\left(F(x,a,b,s), F(x',a,b,s) \right) \leq L_{F_1} d_X(x,x') \ \forall s \in \mathfrak{R}^k, \tag{5.23}$$

for some constants $L_r > 0$ and $L_{F_1} > 0$.

(c) The family of functions $\{F(x,a,b,\cdot); (x,a,b) \in \mathbb{K}\}$ is equi-Lipschitz, i.e., for all $s, s' \in \mathfrak{R}^k$ and $(x,a,b) \in \mathbb{K}$

$$d_X\left(F(x,a,b,s), F(x,a,b,s') \right) \leq L_{F_2} \left| s - s' \right|, \tag{5.24}$$

where $|\cdot|$ is the Euclidean distance in $\mathfrak{R}^k$.

Let $\{v_t\} \subset \mathbb{B}_W(X)$ be the value iteration functions defined as

$$v_0 = 0;$$

and for $t \geq 1$ and $x \in X$,

$$v_t(x) = \min_{b \in B} \max_{a \in A} \left\{ r(x,a,b) + \alpha \int_{\Re^k} v_{t-1}\left[F(x,a,b,s)\right] \theta(ds) \right\}$$

$$= \max_{a \in A} \min_{b \in B} \left\{ r(x,a,b) + \alpha \int_{\Re^k} v_{t-1}\left[F(x,a,b,s)\right] \theta(ds) \right\}.$$

It is easy to prove that

$$\left\| v_t - V_\alpha^\theta \right\|_W \to 0, \text{ as } t \to \infty, \tag{5.25}$$

where V_α^θ is the value of the game. Indeed, from (4.5), for each $x \in X$,

$$\left| v_t(x) - V_\alpha^\theta(x) \right| \leq \sup_{(a,b) \in A \times B} \alpha \int_{\Re^k} \left| v_{t-1}\left[F(x,a,b,s)\right] - V_\alpha^\theta\left[F(x,a,b,s)\right] \right| \theta(ds)$$

$$\leq \gamma_\alpha \left\| v_{t-1} - V_\alpha^\theta \right\|_{\bar{W}} \bar{W}(x),$$

which yields

$$\left\| v_t - V_\alpha^\theta \right\|_{\bar{W}} \leq \gamma_\alpha \left\| v_{t-1} - V_\alpha^\theta \right\|_{\bar{W}}. \tag{5.26}$$

Let $l := \limsup_{t \to \infty} \left\| v_t - V_\alpha^\theta \right\|_{\bar{W}}$. Since $v_t, V_\alpha^\theta \in \mathbb{B}_{\bar{W}}(X)$, we have $l < \infty$. Thus, taking $\limsup$ in (5.26) and using the fact that $\gamma_\alpha < 1$, we obtain $l = 0$, and therefore, from (4.7), (5.25) holds.

Lemma 5.1. *Under Assumption 5.1(a)–(b), the following hold.*
(a) For each $t \in \mathbb{N}_0$, the function v_t is Lipschitz with constant

$$L_{v_t}^\alpha := L_r \sum_{l=0}^{t-1} (\alpha L_{F_1})^l. \tag{5.27}$$

(b) In addition, if $\alpha L_{F_1} < 1$, then the value of the game V_α^θ is Lipschitz with constant

$$L_V^\alpha := \frac{L_r}{1 - \alpha L_{F_1}}.$$

Proof. (a) We proceed by induction. Observe that the part (a) holds for $v_0 = 0$. Now, we assume that v_t is Lipschitz with constant $L_{v_t}^\alpha$ defined in (5.27). Then, for each $x, x' \in X$, from Assumption 5.1 we obtain

$$\left|v_{t+1}(x) - v_{t+1}(x')\right| \leq \sup_{(a,b)\in A\times B} \left\{\left|r(x,a,b) - r(x',a',b')\right|\right.$$

$$+\alpha \int_{\Re^k} \left|v_t\left[F(x,a,b,s)\right] - v_t\left[F(x',a,b,s)\right]\right| \theta(ds)$$

$$\leq L_r d_X(x,x') + \alpha \sup_{(a,b)\in A\times B} L_{v_t}^\alpha \int_{\Re^k} d_X\left(F(x,a,b,s), F(x',a,b,s)\right) \theta(ds)$$

$$\leq L_r d_X(x,x') + \alpha L_{v_t}^\alpha L_{F_1} d_X(x,x')$$

$$= \left[L_r + \alpha L_{F_1} L_r \sum_{l=0}^{t-1} (\alpha L_{F_1})^l\right] d_X(x,x')$$

$$\leq L_r \left[1 + \sum_{l=0}^{t-1} (\alpha L_{F_1})^{l+1}\right] d_X(x,x') = L_r \sum_{l=0}^{t} (\alpha L_{F_1})^l d_X(x,x')$$

$$= L_{v_{t+1}}^\alpha d_X(x,x'). \tag{5.28}$$

This fact proves part (a).

(b) Observe that if $\alpha L_{F_1} < 1$, then

$$\lim_{t\to\infty} L_{v_t}^\alpha = \frac{L_r}{1 - \alpha L_{F_1}} = L_V^\alpha. \tag{5.29}$$

Now, for $x,x' \in X$, by adding and subtracting the terms $v_t(x)$ and $v_t(x')$, we have

$$\left|V_\alpha^\theta(x) - V_\alpha^\theta(x')\right| \leq \left|V_\alpha^\theta(x) - v_t(x)\right| + \left|v_t(x) - v_t(x')\right| + \left|v_t(x') - V_\alpha^\theta(x')\right|$$

$$+ \left|V_\alpha^\theta(x) - v_t(x)\right| + L_{v_t}^\alpha d_X(x,x') + \left|v_t(x') - V_\alpha^\theta(x')\right|. \tag{5.30}$$

Letting $t \to \infty$ in (5.30), from (5.25) and (5.29) we get

$$\left|V_\alpha^\theta(x) - V_\alpha^\theta(x')\right| \leq L_V^\alpha d_X(x,x'),$$

that is, V_α^θ is Lipschitz with constant L_V^α. ∎

The results in Lemma 5.1 together with Assumption 5.1(c) will be used to verify the equicontinuity and equi-Lipschitz conditions in Assumptions 4.7 and 4.9(c).

Observe that, because $W(\cdot) \geq 1$, for all $(x,a,b) \in \mathbb{K}$ and $s, s' \in \mathfrak{R}^k$,

$$\left| \frac{V_\alpha^\theta(F(x,a,b,s))}{W(x)} - \frac{V_\alpha^\theta(F(x,a,b,s'))}{W(x)} \right| \leq \left| V_\alpha^\theta(F(x,a,b,s)) - V_\alpha^\theta(F(x,a,b,s')) \right|$$

$$\leq L_V^\alpha d_X(F(x,a,b,s), F(x,a,b,s'))$$

$$\leq L_V^\alpha L_{F_2} \left| s - s' \right|. \tag{5.31}$$

Hence, the family of function $\mathscr{V}_W$ (see (4.19)) is equi-Lipschitz, and therefore it is equicontinuous.

On the other hand, it is worth noting that if $L_{F_1} < 1$, then, from (5.27), (5.28), and (5.29), the function V_α^θ is Lipschitz with constant

$$L_V^\alpha := \frac{L_r}{1 - L_{F_1}},$$

which is independent on $\alpha \in (0,1)$. Thus, from (5.31), we can prove that the family of functions $\hat{\mathscr{V}}_W$ in Assumption 4.9(c) is equi-Lipschitz.

Example 5.3. We consider the storage game model introduced in Sect. 5.5. In this case the dynamic of the zero-sum game is defined by the function

$$F(x,a,b,s) = (x + aR - bs)^+,$$

where $x \in X := [0,\infty)$, $a \in A := [a_*, 1]$, $b \in B := [b_*, b^*]$, $s \in S := [0,\infty)$, and R is a positive constant. We will prove that F satisfies relations (5.23) and (5.24) in Assumption 5.1.

For arbitrary and fixed $(a,b,s) \in A \times B \times S$, we define the sets

$$K_1 := \{x \in X : x + aR - bs \leq 0\}$$

and

$$K_2 := \{x \in X : x + aR - bs > 0\},$$

and let d_X be the Euclidean distance in $\mathfrak{R}$. If $x, x' \in K_1$, then

$$\left| F(x,a,b) - F(x',a,b) \right| = 0 \leq \left| x - x' \right|, \tag{5.32}$$

and if $x, x' \in K_2$, we have

$$\left| F(x,a,b) - F(x',a,b) \right| = \left| x + aR - bs - x' - aR + bs \right| = \left| x - x' \right|. \tag{5.33}$$

Now, if $x \in K_1$ and $x' \in K_2$, then $F(x,a,b,s) = 0$ and $-(x+aR-bs) \geq 0$. Thus

$$\begin{aligned} \left| F(x,a,b) - F(x',a,b) \right| &= \left| x' + aR - bs \right| \\ &\leq \left| x' + aR - bs - (x+aR-bs) \right| \\ &= \left| x - x' \right|. \end{aligned} \qquad (5.34)$$

Therefore, a combination of (5.32)–(5.34) yields (5.23). ■

Similar arguments show that (5.24) holds with $L_{F_2} := b^*$.

In conclusion, in the cases when Assumptions 4.1 and 4.2 (or Assumptions 3.1 and 3.2) hold together with Assumption 5.1, we can apply the empirical approximation schemes introduced in Chap. 4 for the discounted and the average criteria.

5.7 Numerical Implementations

5.7.1 Linear-Quadratic Games

Consider the LQ-game introduced in Sect. 5.3. We begin our analysis by presenting the solution of the game through explicit formulae of the value function and the optimal strategies. Such expressions will be used to compare the exact solution with the numerical approximations obtained in the implementation of the estimation and control algorithm.

Solution to LQ-Game. We apply the value iteration approach (2.45) in Remark 2.6 considering $\rho_t = \rho \; \forall t \in \mathbb{N}_0$. As is well known, and we will verify it below (see, e.g., [14]), from the quadratic form of the payoff and because the dynamic is additive noise with mean zero, the optimal pair of strategies can be found in the set of pure strategies (see Definition 1.2). In this case, the value iteration functions $\{U_t\}$ take the form $U_0 = 0$, and for $x \in \mathbb{R}$ and $t \in \mathbb{N}$,

$$U_t(x) = \min_{b \in B(x)} \max_{a \in A(x)} \left[x^2 + a^2 - b^2 + \alpha \int_{\mathbb{R}} U_{t-1}(x+a+b+s)\rho(s)ds \right]. \qquad (5.35)$$

Observe that

$$U_1(x) = \min_{b \in B(x)} \max_{a \in A(x)} \left[x^2 + a^2 - b^2 \right], \; x \in \mathbb{R}.$$

Hence, it is easy to see that $f_1^1(x) = 0$, $f_1^2(x) = 0$, and $U_1(x) = x^2$. To obtain U_2, from (5.35) and (5.5) we have, for $x \in \mathbb{R}$,

$$U_2(x) = \min_{b \in B(x)} \max_{a \in A(x)} \left[x^2 + \alpha x^2 + (\alpha + 1)a^2 + (\alpha - 1)b^2 \right.$$

$$\left. + 2\alpha ax + 2\alpha bx + 2\alpha ab \right] + \sigma^2 \alpha. \tag{5.36}$$

To find the saddle point, we compute the partial derivatives of the function within brackets with respect to a, and then with respect to b. Set both derivatives equal to zero and solve for a and b to obtain

$$a = -\alpha x, \quad b = \alpha x.$$

Thus,

$$f_2^1(x) = -\alpha x \text{ and } f_2^2(x) = \alpha x, \quad x \in \mathfrak{R}.$$

Substituting these values in (5.36) we get

$$U_2(x) = (1 + \alpha)x^2 + \sigma^2 \alpha. \tag{5.37}$$

Similarly, letting $t = 3$ in (5.35) and using (5.37), we obtain

$$f_3^1(x) = -(1 + \alpha)x, \quad f_3^2(x) = (1 + \alpha)x,$$

and

$$U_3(x) = (1 + \alpha(1 + \alpha))x^2 + \sigma^2(\alpha^2 + \alpha(1 + \alpha)).$$

In general, proceeding by induction, we obtain the expressions

$$f_t^1(x) = -\beta_{t-1}x, \quad f_t^2(x) = \beta_{t-1}x, \tag{5.38}$$

and

$$U_t(x) = \frac{\beta_t}{\alpha}x^2 + \sigma^2 L_t, \tag{5.39}$$

where $\beta_0 = 0$,

$$\beta_t = \alpha(1 + \beta_{t-1}) = \frac{\alpha - \alpha^{t+1}}{1 - \alpha}, \quad t \in \mathbb{N}, \tag{5.40}$$

and

$$L_t = \sum_{i=1}^{t-1} \alpha^{t-i-1} \beta_i.$$

A straightforward calculation shows that

$$L_t = \frac{\alpha - \alpha^t}{(1 - \alpha)^2} - \frac{(t-1)\alpha^t}{1 - \alpha}. \tag{5.41}$$

In addition, for $\alpha < 1/4$ (see (5.6)) we have that $\beta_t \leq 1/2 \ \forall t \in \mathbb{N}_0$. Then, in this case,

$$f_t^1(x) \in A(x), \quad f_t^2(x) \in B(x),$$

where $A(x) = B(x) = [-|x|/2, |x|/2]$. Furthermore, by observing that

$$\lim_{t \to \infty} \beta_t = \frac{\alpha}{1 - \alpha} \quad \text{and} \quad \lim_{t \to \infty} L_t = \frac{\alpha}{(1 - \alpha)^2},$$

we have

$$\lim_{t \to \infty} U_t(x) = \frac{1}{1 - \alpha} x^2 + \frac{\alpha}{(1 - \alpha)^2} \sigma^2, \quad x \in \mathfrak{R}.$$

Hence, Lemma 2.3 yields (recall $U_t^i = U_t$ for $i = 1, 2$) that the value of the game is

$$V_\alpha^\rho(x) = \frac{1}{1 - \alpha} x^2 + \frac{\alpha}{(1 - \alpha)^2} \sigma^2, \quad x \in \mathfrak{R}. \tag{5.42}$$

In fact, we have the convergence in the W-norm with $W(x) = x^2 + 1$, that is

$$\lim_{t \to \infty} \left\| U_t - V_\alpha^\rho \right\|_W = 0.$$

Finally, we define the selectors

$$f_*^1(x) := \lim_{t \to \infty} f_t^1(x) = \lim_{t \to \infty} -\beta_{t-1} x = -\frac{\alpha}{1 - \alpha} x \in A(x);$$

$$f_*^2(x) := \lim_{t \to \infty} f_t^2(x) = \lim_{t \to \infty} \beta_{t-1} x = \frac{\alpha}{1 - \alpha} x \in B(x).$$

It is easy to prove that the pair (f_*^1, f_*^2) satisfies the relations (2.15) and (2.16) with $V_\alpha^\rho = V_\alpha$. Therefore, from Theorem 2.6 (see Theorem 2.5(b)), $\pi_*^1 = \{f_*^1\}$ and $\pi_*^2 = \{f_*^2\}$ form a stationary optimal pair of strategies for the LQ-game.

Estimation and Control in LQ-Games. As in Sect. 5.3, we assume that $\{\xi_t\}$ is a sequence of i.i.d. random variables normally distributed with mean zero and unknown variance $\sigma^2 \in (0, 3)$, with density ρ given in (5.5). Now, let $\xi_0, \xi_1, \ldots, \xi_{t-1}$ be an i.i.d. sample from the density ρ and $\hat{\sigma}_t^2$ be an estimator of the variance defined as $\hat{\sigma}_1^2 = 0$,

$$\hat{\sigma}_t^2 = \frac{1}{t - 1} \sum_{i=1}^{t} \left(\xi_{i-1} - \bar{\xi} \right)^2, \quad t > 1, \tag{5.43}$$

where $\bar{\xi}$ is the sample mean. Clearly $\hat{\sigma}_t^2$ is an unbiased estimator, that is $E\left[\hat{\sigma}_t^2\right] = \sigma^2$, and an estimator $\hat{\rho}_t$ of the density ρ is

$$\hat{\rho}_t(s) := \frac{1}{\hat{\sigma}_t \sqrt{2\pi}} \exp\left(-\frac{s^2}{2\hat{\sigma}_t^2} \right), \quad s \in \mathfrak{R}, \tag{5.44}$$

where $\hat{\sigma}_t = \sqrt{\hat{\sigma}_t^2}$. Following similar arguments as (5.6) and (5.7), we have that $\hat{\rho}_t \in \mathcal{D}$ (see Definition 2.2 and Remark 2.5). Therefore, from (2.38) we can take $\rho_t(s) = \hat{\rho}_t(s)$. Thus, it is easy to see that

$$E \left\| \rho_t - \rho \right\|_{L_1} \to 0 \quad \text{as } t \to \infty,$$

and moreover (see Lemma 2.1)

$$E \|\rho_t - \rho\| \to 0 \text{ as } t \to \infty.$$

Under this scenario, the value iteration functions (2.45), given in Remark 2.6 (see (5.35)), take the form $U_0 = 0$, and for $t \in \mathbb{N}$,

$$U_t(x) = \min_{b \in B(x)} \max_{a \in A(x)} \left[x^2 + a^2 - b^2 + \alpha \int_{\mathfrak{R}} U_{t-1}(x+a+b+s)\rho_t(s)ds \right].$$

Summarizing, the following algorithm finds a pair of AD-optimal strategies and an approximation to the value of the LQ-game:

Algorithm 5.2 (Estimation and Control) *1. Set* $t = 0$ *and* $U_0 = 0$. *Choose arbitrary* $(\bar{f}_0^1, \bar{f}_0^2) \in \mathbb{F}^1 \times \mathbb{F}^2$, $\alpha < 1/4$.

2. Set $t = 1$. *Find* $(\bar{f}_1^1, \bar{f}_1^2) \in \mathbb{F}^1 \times \mathbb{F}^2$ *such that, for each* $x \in \mathfrak{R}$,

$$U_1(x) = \min_{b \in B(x)} \max_{a \in A(x)} \left[x^2 + a^2 - b^2 \right]$$

$$= \max_{a \in A(x)} \left[x^2 + a^2 - \left[\bar{f}_1^2(x) \right]^2 \right]$$

$$= \min_{b \in B(x)} \left[x^2 + \left[\bar{f}_1^1(x) \right]^2 - b^2 \right].$$

3. For $t > 1$ *and observations* $\xi_1, \xi_2, \ldots, \xi_t$ *from the normal density* ρ, *compute* $\hat{\sigma}_t^2$ *and* $\hat{\rho}_t(s)$ *(see (5.43) and (5.44))*.

4. For each $x \in \mathfrak{R}$ *and* $t > 1$, *let*

$$U_t(x) = \min_{b \in B(x)} \max_{a \in A(x)} \left[x^2 + a^2 - b^2 + \alpha \int_{\mathfrak{R}} U_{t-1}(x+a+b+s)\rho_t(s)ds \right].$$

Find $(\bar{f}_t^1, \bar{f}_t^2) \in \mathbb{F}^1 \times \mathbb{F}^2$ *such that*

$$U_t(x) = \max_{a \in A(x)} \left[x^2 + a^2 - \left[\bar{f}_t^2(x) \right]^2 + \alpha \int_{\mathfrak{R}} U_{t-1}(x+a+\bar{f}_t^2(x)+s)\rho_t(s)ds \right]$$

$$= \min_{b \in B(x)} \left[x^2 + \left[\bar{f}_t^1(x) \right]^2 - b^2 + \alpha \int_{\mathfrak{R}} U_{t-1}(x+\bar{f}_t^1(x)+b+s)\rho_t(s)ds \right].$$

From Theorem 2.11 the strategies $\bar{\pi}_*^1 = \{\bar{f}_t^1\}$ *and* $\bar{\pi}_*^2 = \{\bar{f}_t^2\}$ *are asymptotically discounted optimal, and, from Lemma 2.3,* U_t *approximates the value of the LQ-game* V_α^ρ.

Table 5.1 Sequence of iterates for estimation and control algorithm in LQ-games

t	$U_t(-3)$	$U_t(0)$	$U_t(4)$
0	0.000000	0.000000	0.000000
1	11.16397	0.003969657	19.84397
2	12.30697	0.628568251	21.39017
3	12.48668	0.683862933	21.66665
4	12.40721	0.574538999	21.61041
5	12.31120	0.471361817	21.51997
10	12.48441	0.642306046	21.69493
15	12.72211	0.880008767	21.93264
20	12.87389	0.838517263	22.08441
50	13.02697	0.837747231	21.86044
60	12.95579	0.834927040	21.98144
100	12.77091	0.8321739	21.88293
250	12.67145	0.8311490	21.88478
	$\downarrow$	$\downarrow$	$\downarrow$
	12.67313	0.8310249	21.88366
	$V_\alpha^\rho(-3)$	$V_\alpha^\rho(0)$	$V_\alpha^\rho(4)$

To solve steps 2 and 4 in Algorithm 5.2 we can proceed similarly as (5.38)–(5.41). In this case it is easy to see that

$$\bar{f}_t^1(x) = -\beta_{t-1}x \in A(x), \quad \bar{f}_t^2(x) = \beta_{t-1}x \in B(x),$$

whenever $\alpha < 1/4$, and

$$U_t(x) = \frac{\beta_t}{\alpha}x^2 + \sigma^2 L_t^{\hat{\sigma}_t^2}, \tag{5.45}$$

where

$$L_t^{\hat{\sigma}_t^2} = \sum_{i=1}^{t-1} \alpha^{t-i-1}\beta_i\hat{\sigma}_{i+1}^2. \tag{5.46}$$

Finally, from Lemma 2.3,

$$\lim_{t\to\infty} E_x^{\bar{\pi}_*^1, \bar{\pi}_*^2} \left\| U_t - V_\alpha^\rho \right\|_W = 0.$$

Numerical Results. We fix $\alpha = 0.24$. We are interested in approximating the value function V_α^ρ given in (5.42) for the initial states $x = 0$, $x = -3$, and $x = 4$. Algorithm 5.2 was performed using the expression (5.45) with samples $(\xi_0, \xi_1, \ldots, \xi_{t-1})$ simulated from the normal density (5.5), assuming that the true variance is $\sigma^2 = 2$. The results are given in Table 5.1.

The approximation to the value function V_α^ρ on the entire state space is shown graphically in Fig. 5.1.

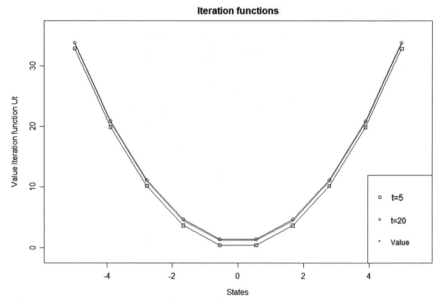

Fig. 5.1 Approximation to the value function V_α^ρ in LQ-games

5.7.2 Finite Games: Empirical Approximation

We now apply the empirical approximation-estimation schemes, introduced in Chap. 4, in a class of finite games. Specifically, based on the results stated in Sect. 4.5, we implement an algorithm defined by the empirical distribution to approximate the value of the game for both discounted and average criteria.

Consider a game evolving according to the equation

$$x_{t+1} = \min\{x_t + a_t + b_t, 3\} + \xi_t, \quad t \in \mathbb{N}_0,$$

where $X = \{3,4,5,6\}$, $A = A(x) = B = B(x) = \{1,2\}$ for all $x \in X$, and $\{\xi_t\}$ is a sequence of discrete i.i.d. random variables with unknown distribution θ, taking values in $S = \{0,1,2,3\}$. The payoff function is given as

$$r(3,a,b) = \begin{cases} 2, & a=1, b=1, \\ -3, & a=1, b=2, \\ -1, & a=2, b=1, \\ 3, & a=2, b=2; \end{cases} \qquad r(4,a,b) = \begin{cases} 2, & a=1, b=1, \\ -1, & a=1, b=2, \\ -1, & a=2, b=1, \\ 1, & a=2, b=2; \end{cases}$$

$$\qquad\qquad\qquad\qquad\qquad\qquad\qquad\qquad\qquad\qquad\qquad\qquad\qquad\qquad (5.47)$$

$$r(5,a,b) = \begin{cases} 4, & a=1, b=1, \\ -2, & a=1, b=2, \\ -2, & a=2, b=1, \\ 2, & a=2, b=2; \end{cases} \qquad r(4,a,b) = \begin{cases} 3, & a=1, b=1, \\ -2, & a=1, b=2, \\ -2, & a=2, b=1, \\ 1, & a=2, b=2. \end{cases}$$

Since the payoff is a bounded function, the W-growth conditions, given in Assumptions 4.1 and 4.3, hold (see Remark 4.1). Furthermore, because the state, control, and disturbance spaces are countable, the continuity and equicontinuity conditions trivially are satisfied. Thus, the results stated in Sect. 4.5 are applicable. However, it is worth noting that it is not possible to ensure, in advance, that the optimal pair belongs to the set of pure strategies. Hence, the analysis will be done on the set of all strategies. In this sense, we consider the sets $\mathbb{A} = \mathbb{P}(A)$ and $\mathbb{B} = \mathbb{P}(B)$ defined as

$$\mathbb{A} := \left\{ \varphi^1 = (\gamma, 1 - \gamma) : \gamma \in [0, 1] \right\}$$

and

$$\mathbb{B} := \left\{ \varphi^2 = (\beta, 1 - \beta) : \beta \in [0, 1] \right\},$$

where γ and $1 - \gamma$ represent the probabilities of selecting actions 1 and 2 by player 1, respectively. Similarly, β and $1 - \beta$ are the probabilities of selecting actions 1 and 2 by player 2.

Let $\bar{r} : X \times [0, 1] \times [0, 1] \to \Re$ be the function defined as

$$\bar{r}(x, \gamma, \beta) = r(x, 1, 1)\gamma\beta + r(x, 1, 2)\gamma(1 - \beta)$$

$$+ r(x, 2, 1)(1 - \gamma)\beta + r(x, 2, 2)(1 - \gamma)(1 - \beta). \tag{5.48}$$

According to (1.5), we have, for each $\varphi^1 = (\gamma, 1 - \gamma) \in \mathbb{A}$ and $\varphi^2 = (\beta, 1 - \beta) \in \mathbb{B}$,

$$r(x, \varphi^1, \varphi^2) = \bar{r}(x, \gamma, \beta), \quad \gamma, \beta \in [0, 1]. \tag{5.49}$$

Discounted Criterion. We apply the value iteration procedure defined in (4.50). Considering (5.49) the functions $\{V_t\}$ take the form $V_0 = 0$, and for $t \in \mathbb{N}$ and $x \in X$,

$$V_t(x) = \max_{\gamma \in [0,1]} \min_{\beta \in [0,1]} \left[\bar{r}(x, \gamma, \beta) + \alpha \int_S \bar{V}_{t-1}[F(x, \gamma, \beta, s)]\theta_t(ds) \right]$$

$$= \max_{\gamma \in [0,1]} \min_{\beta \in [0,1]} \left[\bar{r}(x, \gamma, \beta) + \frac{\alpha}{t} \sum_{i=0}^{t-1} \bar{V}_{t-1}[F(x, \gamma, \beta, s)] \right], \tag{5.50}$$

where

$$\bar{V}_t[F(x, \gamma, \beta, s)] = \bar{V}_t[F(x, 1, 1, s)]\gamma\beta + \bar{V}_t[F(x, 1, 2, s)]\gamma(1 - \beta)$$

$$+ \bar{V}_t[F(x, 2, 1, s)](1 - \gamma)\beta + \bar{V}_t[F(x, 2, 2, s)](1 - \gamma)(1 - \beta).$$

For $t = 1$ and $x = 3$, from (5.47) we have

$$V_1(3) = \max_{\gamma \in [0,1]} \min_{\beta \in [0,1]} [\bar{r}(3, \gamma, \beta)]$$

$$= \max_{\gamma \in [0,1]} \min_{\beta \in [0,1]} \{9\gamma\beta - 6\gamma - 4\beta + 3\}.$$

Computing derivatives of the function within brackets with respect to γ, and then with respect to β, and equaling them to zero, we obtain a system of equations whose solution yields

$$\tilde{\varphi}_1^1(3) = \left(\frac{4}{9},\frac{5}{9}\right), \quad \tilde{\varphi}_1^2(3) = \left(\frac{6}{9},\frac{3}{9}\right).$$

Thus, $V_1(3) = \frac{1}{3}$. We proceed similarly for $x = 4,5,6$, to obtain $V_1(4) = \frac{1}{5}$, $V_1(5) = \frac{2}{5}$, $V_1(6) = -\frac{1}{8}$, and

$$\tilde{\varphi}_1^1(4) = \left(\frac{2}{5},\frac{3}{5}\right), \quad \tilde{\varphi}_1^2(4) = \left(\frac{2}{5},\frac{3}{5}\right),$$

$$\tilde{\varphi}_1^1(5) = \left(\frac{2}{5},\frac{3}{5}\right), \quad \tilde{\varphi}_1^2(5) = \left(\frac{2}{5},\frac{3}{5}\right),$$

$$\tilde{\varphi}_1^1(6) = \left(\frac{3}{8},\frac{5}{8}\right), \quad \tilde{\varphi}_1^2(6) = \left(\frac{3}{8},\frac{5}{8}\right).$$

For $t = 2$ and $x = 3$, from (5.50) we have

$$V_2(3) = \max_{\gamma \in [0,1]} \min_{\beta \in [0,1]} \left[\bar{r}(x,\gamma,\beta) + \frac{\alpha}{2} \sum_{i=0}^{1} \bar{V}_1[F(x,\gamma,\beta,s)] \right]$$

$$= \max_{\gamma \in [0,1]} \min_{\beta \in [0,1]} \left[9\gamma\beta - 6\gamma - 4\beta + 3 + \frac{\alpha}{2}[V_1(3+\xi_0) + V_1(3+\xi_1)] \right].$$

In addition $\tilde{\varphi}_2^1(3) = \left(\frac{4}{9},\frac{5}{9}\right)$ and $\tilde{\varphi}_2^2(3) = \left(\frac{6}{9},\frac{3}{9}\right)$. In fact, for $x = 3,4,5,6$ we can prove that

$$V_2(x) = V_1(x) + \frac{\alpha}{2} \sum_{i=0}^{1} V_1(3+\xi_i).$$

In general, a straightforward calculation shows that, for $x \in X$ and $t \in \mathbb{N}$

$$V_t(x) = V_{t-1}(x) + \frac{\alpha}{t} \sum_{i=0}^{t-1} V_{t-1}(3+\xi_i). \tag{5.51}$$

Numerical Results. We implement the empirical approximation scheme by taking samples $(\xi_0, \xi_1, \ldots, \xi_{t-1})$ from the Binomial distribution θ with parameters $(3, 1/2)$, that is

$$P[\xi_t = k] = \theta(k) = \binom{3}{k} \left(\frac{1}{2}\right)^3, \quad k = 0,1,2,3.$$

It is worth noting that if we take $\theta_t = \theta$ as the true distribution in (5.50), from (5.51) it is easy to see that

Table 5.2 Sequence of iterates for empirical value iteration

t	$V_t(3)$	$V_t(4)$	$V_t(5)$	$V_t(6)$
0	0.000000	0.000000	0.000000	0.000000
1	0.3333333	0.2000000	0.4000000	-0.12500000
2	0.5166667	0.3833333	0.5833333	0.05833333
3	0.6138889	0.4805556	0.6805556	0.15555556
4	0.6402778	0.5069444	0.7069444	0.18194444
5	0.6076389	0.4743056	0.6743056	0.14930556
10	0.5534278	0.4200945	0.6200945	0.09509445
15	0.5638499	0.4305166	0.6305166	0.10551658
20	0.5569752	0.4236419	0.6236419	0.09864192
50	0.5597872	0.4264539	0.6264539	0.10145386
60	0.5649057	0.4315723	0.6315723	0.10657232
100	0.5844444	0.4511110	0.6511110	0.12611104
250	0.5885507	0.4552173	0.6552173	0.12621734
	$\downarrow$	$\downarrow$	$\downarrow$	$\downarrow$
	0.5843750	0.4510417	0.6510417	0.1260416667
	$V_\alpha^\theta(3)$	$V_\alpha^\theta(4)$	$V_\alpha^\theta(5)$	$V_\alpha^\theta(6)$

$$V_t(x) = V_{t-1}(x) + \alpha \sum_{i=3}^{6} V_{t-1}(i)\theta(i-3)$$

$$= V_1(x) + 0.2510417 \left(\frac{\alpha - \alpha^t}{1 - \alpha} \right).$$

Hence, letting $t \to \infty$ we obtain the value of the game

$$V_\alpha^\theta(x) = V_1(x) + \frac{0.2510417}{1-\alpha}\alpha, \quad x = 3,4,5,6. \tag{5.52}$$

This expression will be used to compare the numerical approximation with the exact solution.

The numerical results shown in Table 5.2 correspond to the value $\alpha = 1/2$. Specifically, we show the approximation to the exact values $V_\alpha^\theta(3) = 0.584375$, $V_\alpha^\theta(4) = 0.4510417$, $V_\alpha^\theta(5) = 0.6510417$, and $V_\alpha^\theta(6) = 0.1260417$, obtained from (5.52).

In Fig. 5.2 we show the approximation to the value function V_α^θ for different samples $(\xi_0, \xi_1, \dots, \xi_{t-1})$ from the Binomial distribution.

Average Criterion. We apply the empirical VDFA introduced in Sect. 4.5 to approximate the value j^* of the average game. Observe that from (5.52) and (4.13), for any sequence of discount factors $\alpha_t \to 1$ we have,

$$j^* = \lim_{t \to \infty} j_{\alpha_t}^\theta = \lim_{t \to \infty} (1 - \alpha_t) V_{\alpha_t}^\theta(z) = 0.2510417, \quad \forall z = 3,4,5,6.$$

As is stated in Sect. 4.5, in order to obtain an approximation to the value $j^* = 0.2510417$, we fix a suitable sequence $\{\bar{\alpha}_t\}$ converging, slowly enough, to one.

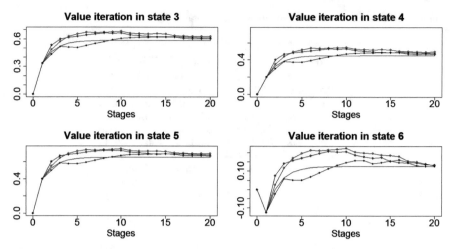

Fig. 5.2 Simulations of the empirical value iteration functions for different samples from the Binomial distribution

Table 5.3 Empirical approximation to the average value function

t	$\bar{\alpha}_t$	$j_n^{(t)}(3)$	$j_n^{(t)}(4)$	$j_n^{(t)}(5)$	$j_n^{(t)}(6)$
0	0.2928932	0.000000	0.000000	0.000000	0.000000
1	0.4226497	1.0000000	1.0000000	1.0000000	1.00000000
2	0.5000000	0.2933915	0.2267248	0.3267248	0.06422481
3	0.5527864	0.2891668	0.2295384	0.3189811	0.08419393
4	0.5917517	0.2860465	0.2316134	0.3132631	0.09893273
5	0.6220355	0.2836202	0.2332249	0.3088178	0.13929619
10	0.7113249	0.2764585	0.2365061	0.2957035	0.14843031
15	0.7574644	0.2727517	0.2400184	0.2889207	0.16417203
20	0.7867904	0.2703918	0.2416999	0.2846052	0.17444542
50	0.8613250	0.2644850	0.2458146	0.2738202	0.20030550
60	0.8729999	0.2635119	0.2464404	0.2720477	0.20521931
100	0.9009852	0.2611855	0.2479183	0.2678191	0.21557962
250	0.9370059	0.2581822	0.2497663	0.2623902	0.22925250
500	0.9553678	0.2566325	0.2506756	0.2596110	0.23615570
1000	0.9684088	0.2555126	0.2512983	0.2576197	0.24102606
1500	0.9741973	0.2550064	0.2515661	0.2567266	0.24318020
	$\downarrow$	$\downarrow$	$\downarrow$	$\downarrow$	
	1.0000000	0.2510417	0.2510417	0.2510417	0.2510417
		j^*	j^*	j^*	j^*

Then, for each $\bar{\alpha}_t$, we apply the approximation scheme given by (4.52), where the functions $V_n^{(t)}$ can be obtained from (5.51) with $\bar{\alpha}_t$ instead of α. Hence, from (4.56), $j_n^{(t)}(z) := (1 - \bar{\alpha}_t)V_n^{(t)}(z)$ is an estimator of j^*.

Table 5.3 shows the approximation to the constant $j^* = 0.2510417$ taking the sequence $\bar{\alpha}_t = 1 - \dfrac{1}{\sqrt{t+1}}$. From a practical point of view, the algorithm was performed letting $t = n$. It is worth noting that, really, for any $z \in \{3,4,5,6\}$, we obtain the approximation to the value of the average game j^*.

Appendix A
Elements from Analysis

A.1 Semicontinuous Functions

Unless stated otherwise, throughout the following we suppose that X is a topological space, and we denote the extended real numbers set as $\bar{\mathfrak{R}}$, i.e., $\bar{\mathfrak{R}} = \mathfrak{R} \cup \{-\infty, \infty\}$.

Definition A.1. A function $v : X \to \bar{\mathfrak{R}}$ is said to be
   (a) lower semicontinuous (l.s.c.) if the set $\{x \in X : v(x) \leq r\}$ is closed in X for every $r \in \mathfrak{R}$;
   (b) upper semicontinuous (u.s.c.) if the set $\{x \in X : v(x) \geq r\}$ is closed in X for every $r \in \mathfrak{R}$.

   Clearly, a function v is l.s.c. if and only if $-v$ is u.s.c., and v is continuous if and only if v is l.s.c. and u.s.c.

   The next results summarize the main properties of semicontinuous functions. For their proofs, see, e.g., [3, 5].

Proposition A.1. *Let X be a metric space and $v : X \to \bar{\mathfrak{R}}$. Then:*
   (a) v is l.s.c. if and only if for each sequence $\{x_n\}$ in X such that $x_n \to x \in X$, as $n \to \infty$, we have $\liminf_{n \to \infty} v(x_n) \geq v(x)$.
   (b) v is l.s.c. and bounded from below if and only if there exists a sequence $\{v_n\}$ of bounded and continuous functions such that $v_n \nearrow v$.

Remark A.1. Similarly, v is u.s.c. if and only if for each sequence $\{x_n\}$ in X such that $x_n \to x \in X$, as $n \to \infty$, we have $\limsup_{n \to \infty} v(x_n) \leq v(x)$. Moreover, v is u.s.c. and bounded from above if and only if there exists a sequence $\{v_n\}$ of bounded and continuous functions such that $v_n \searrow v$.

© The Author(s), under exclusive license to Springer Nature Switzerland AG 2020
J. A. Minjárez-Sosa, *Zero-Sum Discrete-Time Markov Games*
with Unknown Disturbance Distribution, SpringerBriefs in Probability
and Mathematical Statistics, https://doi.org/10.1007/978-3-030-35720-7

Proposition A.2. *Let X be a compact metric space.*

(a) If $v : X \to \tilde{\mathfrak{R}}$ is l.s.c., then v attains its infimum, i.e., there exists $x_* \in X$ such that $v(x_*) = \inf_{x \in X} v(x)$.

(b) If $v : X \to \tilde{\mathfrak{R}}$ is u.s.c., then v attains its supremum, i.e., there exists $x^* \in X$ such that $v(x^*) = \sup_{x \in X} v(x)$.

A topological space X is a *Hausdorff space* if for each $x, y \in X$, with $x \neq y$, there exist neighborhoods U_x and V_y of x and y, respectively, such that $U_x \cap V_y = \emptyset$. For instance, all metric spaces are Hausdorff.

Theorem A.1 (Fan Minimax Theorem). *Let X and Y be two compact Hausdorff space and $h : X \times Y \to \mathfrak{R}$ be a real valued function. Suppose that*

(a) $h(x, y)$ is u.s.c. in $x \in X$ for each $y \in Y$, and l.s.c. in $y \in Y$ for each $x \in X$;

(b) h is concave in X and convex in Y.

Then the following equality holds:

$$\max_{x \in X} \min_{y \in Y} h(x, y) = \min_{y \in Y} \max_{x \in X} h(x, y).$$

Proof. See [15].

A.2 Spaces of Functions

For a function $\rho : \mathfrak{R}^k \to \mathfrak{R}$ and $1 \leq q < \infty$, we define the L_q-norm

$$\|\rho\|_{L_q} := \left(\int_{\mathfrak{R}^k} |\rho|^q \, d\mu \right)^{1/q},$$

where μ is the Lebesgue measure on $\mathfrak{R}^k$. The space $L_q = L_q(\mathfrak{R}^k)$ consists of all real-valued measurable functions on $\mathfrak{R}^k$ with finite L_q-norm:

$$L_q = \left\{ \rho : \|\rho\|_{L_q} < \infty \right\}.$$

A *Borel space* is a Borel subset of a complete separable metric space. A Borel space is always endowed with the Borel σ-algebra $\mathscr{B}(X)$, that is, the smallest σ-algebra of subsets of X that contains all the open sets in X. In this sense, "measurable," for either sets or functions, means "Borel measurable."

For a Borel space X, we define the following spaces:

- $\mathbb{B}(X)$, Banach space of real-valued bounded measurable functions on X with the supremum norm
$$\|v\|_B := \sup_{x \in X} |v(x)|.$$

- $\mathbb{C}(X) \subset \mathbb{B}(X)$, subspace of bounded continuous functions.
- $\mathbb{L}(X)$, space of l.s.c. functions that are bounded from below.

Let $W : X \to [1,\infty)$ be a measurable function. We define:

- $\mathbb{B}_W(X)$, normed linear space of measurable functions with finite weighted norm (W-norm)

$$\|v\|_W := \left\| \frac{v}{W} \right\|_B = \sup_{x \in X} \frac{|v(x)|}{W(x)}. \tag{A.1}$$

We will refer to W as a weight function. Observe that $\|v\|_W \le \|v\|_B < \infty$ for all $v \in \mathbb{B}(X)$; that is, a bounded function is a W-bounded function. Moreover, $\mathbb{B}_W(X)$ is a Banach space that contains $\mathbb{B}(X)$ (see [31]).

We define the spaces of functions:

$$\mathbb{C}_W(X) = \mathbb{C}(X) \cap \mathbb{B}_W(X) \subset \mathbb{B}_W(X)$$

and

$$\mathbb{L}_W(X) = \mathbb{L}(X) \cap \mathbb{B}_W(X) \subset \mathbb{B}_W(X).$$

That is, $\mathbb{C}_W(X)$ is the subspace of W-bounded continuous functions whereas $\mathbb{L}_W(X)$ is the subspace of W-bounded and l.s.c. functions.

A.3 Multifunctions and Selectors

Throughout the following, X and Y are Borel spaces.

Definition A.2. A multifunction Ψ from X to Y is a function such that $\Psi(x)$ is a nonempty subset of Y for all $x \in X$.

A multifunction is also called a correspondence or set-valued mapping. We use the notation $\Psi : X \twoheadrightarrow Y$ for a multifunction Ψ from X to Y.

A multifunction $\Psi : X \twoheadrightarrow Y$ is said to be compact-valued (respectively, closed-valued) if for each $x \in X$, $\Psi(x)$ is a compact (resp. closed) subset of Y. Its graph is the subset $Gr(\Psi) \subset X \times Y$ defined as

$$Gr(\Psi) := \{(x,y) \in X \times Y : y \in \Psi(x)\}.$$

If Y_0 is a nonempty subset of Y, we define $\Psi^{-1}[Y_0] := \{x \in X : \Psi(x) \cap Y_0 \ne \emptyset\}$.

Definition A.3. Let $\Psi : X \twoheadrightarrow Y$ be a multifunction. It is said that:

(a) Ψ is Borel-measurable if $\Psi^{-1}[G]$ is a Borel subset of X for each open subset $G \subset Y$;

(b) Ψ is u.s.c. if $\Psi^{-1}[F]$ is closed in X for each closed subset $F \subset Y$;

(c) Ψ is l.s.c. if $\Psi^{-1}[G]$ is open in X for each open subset $G \subset Y$;

(d) Ψ is continuous if it is upper and lower semicontinuous.

Proposition A.3. *Let $\Psi : X \twoheadrightarrow Y$ be a compact-valued multifunction. The following statements are equivalent:*

(a) Ψ is Borel-measurable.

(b) $\Psi^{-1}[F]$ is a Borel subset of X for each closed subset $F \subset Y$.

(c) $Gr(\Psi)$ is a Borel subset of $X \times Y$.

(d) Ψ is a measurable function from X to the space of nonempty compact subset of Y topologized by the Hausdorff metric.

Proof. See, e.g., [37, 66].

Definition A.4. Let $\Psi : X \twoheadrightarrow Y$ be a Borel-measurable multifunction. A measurable function $f : X \to Y$ such that $f(x) \in \Psi(x), x \in X$, is called a measurable selector for Ψ.

A measurable selector for Ψ is also called decision function for Ψ, and results stating its existence are called *measurable selection theorems*.

Theorem A.2. *Let $\Psi : X \twoheadrightarrow Y$ be a Borel-measurable compact-valued multifunction and $v : Gr(\Psi) \to \Re$ be a measurable function.*

(a) If $v(x, \cdot)$ is u.s.c. on $\Psi(x)$ for each $x \in X$, then there exists a measurable selector f^* of Ψ such that

$$v(x, f^*(x)) = \max_{y \in \Psi(x)} v(x, y) =: v^*(x) \quad \forall x \in X,$$

and v^* is measurable.

(b) If $v(x, \cdot)$ is l.s.c. on $\Psi(x)$ for each $x \in X$, then there exists a measurable selector f_* of Ψ such that

$$v(x, f_*(x)) = \min_{y \in \Psi(x)} v(x, y) =: v_*(x) \quad \forall x \in X,$$

and v_* is measurable.

Proof. See, e.g., [37, 66].

Proposition A.4. *Let $\Psi : X \twoheadrightarrow Y$ be a Borel-measurable compact-valued multifunction. If $\{f_n\}$ is a sequence of measurable selectors for Ψ, then there exists a measurable selector f^* for Ψ such that $f^*(x) \in \Psi(x)$ is an accumulation point of the sequence $\{f_n(x)\}$ for each $x \in X$.*

Proof. See, e.g., [66].

Theorem A.3 (Berge Maximum Theorem). *Let* $\Psi : X \twoheadrightarrow Y$ *be a continuous compact-valued multifunction between topological spaces* X *and* Y, *and* $v : Gr(\Psi) \to \Re$ *be a continuous function. Define the multifunction* $\Psi^* : X \twoheadrightarrow Y$ *by*

$$\Psi^*(x) = \{y \in \Psi(x) : v(x,y) = v^*(x)\},$$

where $v^*(x) = \max_{y \in \Psi(x)} v(x,y)$. *Then:*

(a) v^* *is continuous.*

(b) Ψ^* *is a compact-valued multifunction.*

(c) If either v *has a continuous extension to* $X \times Y$ *or* Y *is Hausdorff,* Ψ^* *is an u.s.c. multifunction.*

Proof. See [4, pp. 115-116] (see also [1]).

Appendix B
Probability Measures and Weak Convergence

Throughout the following, X is a Borel space with Borel σ-algebra $\mathscr{B}(X)$.

Definition B.1. Let $\mu, \mu_t, t \in \mathbb{N}$, probability measures on X. It is said that μ_t converges weakly to μ (denoted as $\mu_t \overset{w}{\to} \mu$) if

$$\int v d\mu_t \to \int v d\mu$$

for all $v \in \mathbb{C}(X)$.

We denote by $\mathbb{P}(X)$ the space of all probability measures on X. We assume that $\mathbb{P}(X)$ is endowed with the topology of weak convergence given in Definition B.1. In this case, as X is a Borel space, $\mathbb{P}(X)$ is a Borel space (see, e.g., [38, p. 91], [5, Cor. 7.25.1]). In addition, if X is compact, then so is $\mathbb{P}(X)$ (see, e.g., [62, Th. 6.4]). Furthermore, let

$$\mathbb{P}_0(X) := \left\{ \mu \in \mathbb{P}(X) : \int_X |s| d\mu(s) < \infty \right\}.$$

From [49], if X is a σ-compact set, then so is $\mathbb{P}_0(X)$.

Proposition B.1. *For* $\mu, \mu_t \in \mathbb{P}(X), t \in \mathbb{N}$, *the following statements are equivalent:*
(a) $\mu_t \overset{w}{\to} \mu$.
(b) $\liminf\limits_{t \to \infty} \mu_t(D) \geq \mu(D)$ *for every open set* $D \subset X$ *and* $\mu_t(X) \to \mu(X)$.
(c) $\limsup\limits_{t \to \infty} \mu_t(D) \leq \mu(D)$ *for every closed set* $D \subset X$ *and* $\mu_t(X) \to \mu(X)$.

Proof. See, e.g., [3].

© The Author(s), under exclusive license to Springer Nature Switzerland AG 2020
J. A. Minjárez-Sosa, *Zero-Sum Discrete-Time Markov Games
with Unknown Disturbance Distribution*, SpringerBriefs in Probability
and Mathematical Statistics, https://doi.org/10.1007/978-3-030-35720-7

Remark B.1. From Proposition A.1, it is easy to prove that if $\mu_t \overset{w}{\to} \mu$ and $v \in \mathbb{L}(X)$, then

$$\liminf_{t \to \infty} \int v d\mu_t \geq \int v d\mu.$$

Similarly, from Remark A.1, if $\mu_t \overset{w}{\to} \mu$ and v is u.s.c. and bounded from above, then

$$\limsup_{t \to \infty} \int v d\mu_t \leq \int v d\mu.$$

For a function $u : X \to \mathfrak{R}$, we define the mapping $\tilde{u} : \mathbb{P}(X) \to \mathfrak{R}$ as $\tilde{u}(\mu) = \int u d\mu$.

Proposition B.2. *(a) If $u \in \mathbb{L}(X)$, then $\tilde{u} \in \mathbb{L}(\mathbb{P}(X))$;*
(b) if u is u.s.c. and bounded from above, then $\tilde{u}$ is u.s.c. and bounded from above on $\mathbb{P}(X)$.

Proof. See, e.g., [21, Lemma 3.3].

Proposition B.3. *For a multifunction $\Psi : X \twoheadrightarrow Y$, let $\bar{\Psi} : X \twoheadrightarrow \mathbb{P}(X)$ be the multifunction defined as $\bar{\Psi}(x) := \mathbb{P}(\Psi(x))$. If Ψ is continuous, then so is $\bar{\Psi}$.*

Proof. See, e.g., [36, Theorem 3].

Let $\mathscr{V}$ be a family of real-valued measurable function defined on X.

Definition B.2. Let $\mu \in \mathbb{P}(X)$. It is said that $\mathscr{V}$ is a μ-uniformity class if

$$\sup_{v \in \mathscr{V}} \left| \int v d\mu_t - \int v d\mu \right| \to 0 \quad \text{as } t \to \infty,$$

for any sequence $\{\mu_t\} \subset \mathbb{P}(X)$ such that $\mu_t \overset{w}{\to} \mu$.

The following result states the μ-uniformity for equicontinuous families of functions. A family $\mathscr{V}$ is equicontinuous at each $x \in X$, if for each $x \in X$ and $\varepsilon > 0$, there exists $\delta > 0$ such that

$$d(x, x') < \delta, \ x' \in X, \ \text{implies} \ |v(x) - v(x')| < \varepsilon \ \text{for all } v \in \mathscr{V},$$

where d is the metric on X.

Proposition B.4. *If $\mathscr{V}$ is uniformly bounded and equicontinuous at each $x \in X$, then $\mathscr{V}$ is a μ-uniformity class for every probability measure $\mu \in \mathbb{P}(X)$.*

Proof. See, e.g., [64, Theorem 3.1] (see also [6]).

B.1 The Empirical Distribution

Let ξ be a random variable defined on a probability space $(\Omega, \mathscr{F}, P)$ taking values in a Borel space S. We denote by $\theta \in \mathbb{P}(S)$ the distribution of ξ, that is, for $D \in \mathscr{B}(S)$,

$$\theta(D) := P[\xi \in D].$$

Throughout the remainder, $(\Omega, \mathscr{F}, P)$ is a fixed probability space and "a.s." means "almost surely with respect to P."

Let S be a Borel space and $\bar{\xi}_t = (\xi_1, \xi_2, \ldots, \xi_t)$ be an independent and identically distributed (i.i.d.) sample from the distribution $\theta \in \mathbb{P}(S)$. The *empirical distribution* θ_t of the sample $\bar{\xi}_t$ is defined as

$$\theta_t(D) := \frac{1}{t} \sum_{i=1}^{t} 1_D(\xi_i), \quad t \in \mathbb{N}, D \in \mathscr{B}(S).$$

Observe that θ_t is a (random) probability measure that puts mass $1/t$ at each observation, and for a function $v : S \to \mathfrak{R}$

$$\int_S v \, d\theta_t = \frac{1}{t} \sum_{i=1}^{t} v(\xi_i), \quad t \in \mathbb{N}.$$

Furthermore, since, for each $D \in \mathscr{B}(S)$, $\{1_D(\xi_i)\}$ is a sequence of i.i.d. random variables with mean $E[1_D(\xi_i)] = \theta(D)$, from the Strong Law of Large Numbers (SLLN)

$$\theta_t(D) \to \theta(D) \quad \text{a.s., as } t \to \infty.$$

Remark B.2. In the particular case when $S = \mathfrak{R}$, the study of the empirical distribution focuses on the *empirical distribution function*. Specifically, if $\xi_1, \xi_2, \ldots, \xi_t$ are i.i.d. random variables with distribution function F, we define the empirical distribution function as

$$F_t(s) = \theta_t((-\infty, s]), \quad s \in \mathfrak{R}.$$

Thus, from the SLLN we have that $F_t(s) \to F(s)$ a.s., $s \in \mathfrak{R}$. In fact, we have uniform convergence, that is

$$\sup_{s \in \mathfrak{R}} |F_t(s) - F(s)| \to 0 \quad \text{a.s.}$$

This result is known as Glivenko-Cantelli Theorem (see, e.g., [18]).

A well-known property of the empirical distribution is the weak convergence given in the following result (see, e.g., [18]).

Proposition B.5. $\theta_t \overset{w}{\to} \theta$ *a.s., that is,*

$$\int v \, d\theta_t \to \int v \, d\theta \quad \text{a.s., as } t \to \infty,$$

for every $v \in \mathbb{C}(S)$. In addition, if $v \in \mathbb{L}(S)$, then (see Remark B.1)

$$\liminf_{t\to\infty} \int u d\theta_t \geq \int u d\theta \ \ a.s.$$

Similarly if v is only u.s.c. and bounded from above

$$\limsup_{t\to\infty} \int v d\theta_t \leq \int v d\theta.$$

A combination of the Propositions B.4 and B.5 yields the following result (see [64, Theorem 3.1]).

Proposition B.6. *Let $\mathcal{V}$ be a family of real-valued function defined on S. If $\mathcal{V}$ is uniformly bounded and equicontinuous at each $s \in S$, then*

$$\eta_t := \sup_{v \in \mathcal{V}} \left| \int v d\theta_t - \int v d\theta \right| \to 0 \ \ a.s., \ as \ t \to \infty. \tag{B.1}$$

A family of function satisfying (B.1) is called a *Glivenko-Cantelli class*. This class of functions is widely studied in the field of statistical learning theory (see, e.g., [71]).

We now present an important result that provides the rate of convergence of the expectation $E\eta_t$ for a particular family of functions.

Proposition B.7. *Suppose that $S = \Re^k$ and*

$$\int |s|^{\bar{m}} d\theta < \infty \ for \ \bar{m} = \frac{km}{(m-k)(m-2)} \ and \ some \ m > \max\{2,k\}.$$

If $\mathcal{V}$ is uniformly bounded and equi-Lipschitzian on $\Re^k$, then there is a constant $\bar{M}$ such that

$$E\eta_t \leq \bar{M} t^{-1/m}.$$

Proof. See [12, Theorem 3.2 and Proposition 3.4].

The family $\mathcal{V}$ equi-Lipschitzian on $\Re^k$ means that there exists a constant $L > 0$ such that, for every $s, s' \in \Re^k$ and $v \in \mathcal{V}$

$$|v(s) - v(s')| \leq L|s - s'|,$$

where $|\cdot|$ is the Euclidean distance in $\Re^k$.

Appendix C
Stochastic Kernels

Throughout the following, X and Y are Borel spaces.

Definition C.1. A stochastic kernel $\gamma(dx|y)$ on X given Y is a function such that $\gamma(\cdot|y) \in \mathbb{P}(X)$ for every $y \in Y$ and $\gamma(D|\cdot) \in \mathbb{B}(Y)$ for each fixed $D \in \mathscr{B}(X)$.

We denote by $\mathbb{P}(X|Y)$ the family of stochastic kernels on X given Y.

Definition C.2. Let $\gamma \in \mathbb{P}(X|Y)$. It said that
(a) γ is strongly continuous if the function

$$y \to \int v(x)\gamma(dx|y) \tag{C.1}$$

is bounded and continuous for each function $v \in \mathbb{B}(X)$.
(b) γ is weakly continuous if the function in (C.1) is bounded and continuous for each $v \in \mathbb{C}(X)$.

The following result establishes some of the main properties of the stochastic kernels.

Proposition C.1. *Let* $\gamma \in \mathbb{P}(X|Y)$.
(a) The following statements are equivalent:
   (a.1) γ is strongly continuous.
   (a.2) The function in (C.1) is l.s.c. for each $v \in \mathbb{B}(X)$.
   (a.3) $\gamma(D|\cdot)$ is continuous on Y for every $D \in \mathscr{B}(X)$.
(b) The following statements are equivalent:
   (b.1) γ is weakly continuous.
   (b.2) The function in (C.1) is l.s.c. for each $v \in \mathbb{L}(X)$.

Proof. See, e.g., [30, Appendix C].

© The Author(s), under exclusive license to Springer Nature Switzerland AG 2020
J. A. Minjárez-Sosa, *Zero-Sum Discrete-Time Markov Games*
with Unknown Disturbance Distribution, SpringerBriefs in Probability
and Mathematical Statistics, https://doi.org/10.1007/978-3-030-35720-7

Proposition C.2 (Theorem of C. Ionescu Tulcea). *Let $\{X_t\}_{t\in\mathbb{N}_0}$ be a sequence of Borel spaces, $Y_t := X_0 \times X_1 \times \ldots \times X_t$, $t \in \mathbb{N}_0$, and $Y := \prod_{t=0}^{\infty} X_t$. In addition, let $\nu \in \mathbb{P}(X_0)$ be an arbitrary probability measure, and for $t \in \mathbb{N}_0$, $\gamma_t \in \mathbb{P}(X_{t+1}|Y_t)$. Then there exists a unique probability measure P_ν on Y such that, for every measurable rectangle $D_0 \times D_1 \times \ldots \times D_t$ in Y_t*

$$P_\nu(D_0 \times \ldots \times D_t) = \int_{D_0} \nu(dx_0) \int_{D_1} \gamma_0(dx_1|y_0) \int_{D_2} \gamma_1(dx_2|y_1) \ldots \int_{D_t} \gamma_{t-1}(dx_t|y_{t-1}),$$

where $y_t = (x_0, x_1, \ldots, x_t) \in Y_t$.

Proof. See, e.g., [5, pp. 140-141]. ∎

C.1 Difference-Equation Processes

Let $\{x_t\}$ be a stochastic process on X defined by

$$x_{t+1} = F(x_t, \xi_t), \quad t \in \mathbb{N}_0, \quad x_0 \in X \text{ given}, \tag{C.2}$$

where $\{\xi_t\}$ is a sequence of i.i.d. random variables taking values on a Borel space S, with a common distribution $\theta \in \mathbb{P}(S)$, and independent of the initial state x_0. In addition, $F : X \times S \to X$ is a given measurable function.

Equation (C.2), together with the distribution θ, defines a stochastic kernel $\gamma \in \mathbb{P}(X|X)$ by

$$\begin{aligned}
\gamma(D|x) : &= \Pr[x_{t+1} \in D | x_t = x] \\
&= \Pr[F(x, \xi_t) \in D | x_t = x] \\
&= \theta\{s \in S : F(x, s) \in D\} \\
&= \int_S 1_D[F(x, a)]\,\theta(ds), \quad D \in \mathscr{B}(X),\ x \in X. \tag{C.3}
\end{aligned}$$

Hence, for any measurable function v on X

$$E[v(x_{t+1})|x_t = x] = \int_X v(y)\gamma(dy|x) = \int_S v[F(x,a)]\,\theta(ds), \quad x \in X,$$

whenever the integrals exist.

In particular, if $S = \mathfrak{R}^k$ and θ has a density ρ with respect to the Lebesgue measure, i.e., $\theta(D) = \int_D \rho(s)ds$ for all $D \in \mathscr{B}(\mathfrak{R}^k)$, the stochastic kernel γ in (C.3) becomes

$$\gamma(D|x) = \int_{\mathfrak{R}^k} 1_D[F(x,a)]\rho(s)ds, \quad D \in \mathscr{B}(X),\ x \in X, \tag{C.4}$$

and

$$\int_X v(y)\gamma(dy|x) = \int_{\Re^k} v[F(x,a)]\rho(s)ds, \ \ x \in X.$$

Proposition C.3. *If the function $F(x,s)$ in (C.2) is continuous in $x \in X$ for each $s \in S$, then the stochastic kernel γ in (C.3) is weakly continuous.*

Proof. See Example C.7 in [30].

Proposition C.4. *Assume that $X = S = \Re^k$, $F(x,s) = G(x) + s$ for some measurable function G, and θ has a density ρ with respect to the Lebesgue measure. Then the stochastic kernel γ in (C.3) (see (C.4)) is strongly continuous.*

Proof. See Example C.8 in [30].

Appendix D
Review on Density Estimation

We consider random variables ξ defined on a probability space $(\Omega, \mathscr{F}, P)$ taking values in $S = \mathfrak{R}^k$ with density function ρ, that is, for all $D \in \mathscr{B}(\mathfrak{R}^k)$,

$$\theta(D) = \int_D \rho(s)ds,$$

where $\theta \in \mathbb{P}(\mathfrak{R}^k)$ is the distribution of ξ, i.e., $\theta(D) := P[\xi \in D]$; we denote by E the corresponding expectation operator.

Let $\bar{\xi}_t := (\xi_1, \xi_2, \ldots, \xi_t)$ be an i.i.d. sample from a density ρ. A *density estimate* is a sequence $\{\hat{\rho}_t\}$ of Borel measurable functions $\hat{\rho}_t : \mathfrak{R}^k \times (\mathfrak{R}^k)^t \to \mathfrak{R}$ such that, for each $t \in \mathbb{N}$ and sample $\bar{\xi}_t$, $\hat{\rho}_t(s; \bar{\xi}_t) = \hat{\rho}_t(s)$ is a density on $\mathfrak{R}^k$.

D.1 Error Criteria

The objective of *density estimation methods* is to obtain an estimate $\hat{\rho}_t$ under the least restrictive conditions possible on ρ. Hence, the density estimation problem can be considered as a functional approximation problem for which techniques from function approximation theory could be applied. From this point of view, there are several error criteria to measure the discrepancy between the estimate $\hat{\rho}_t$ and the density ρ. For instance, the L_q-criteria, $q \geq 1$, defined by the L_q-norm

$$\left(\int |\hat{\rho}_t(s) - \rho(s)|^q \, ds \right)^{1/q},$$

being the most common cases $q = 1$ and $q = 2$. Specifically we define:

- the integrated absolute error and the mean integrated absolute error

$$J_t := \int |\hat{\rho}_t(s) - \rho(s)| ds = \|\hat{\rho}_t - \rho\|_{L_1}$$

and

$$E(J_t) = E \|\hat{\rho}_t - \rho\|_{L_1},$$

respectively;
- the integrated squared error and the mean integrated squared error

$$J_t^{(2)} := \int (\hat{\rho}_t(s) - \rho(s))^2 ds = \|\hat{\rho}_t - \rho\|_{L_2}^2$$

and

$$E\left(J_t^{(2)}\right) = E \|\hat{\rho}_t - \rho\|_{L_2}^2,$$

respectively.

From these two criteria, the squared error is the most studied because it is more tractable and allows analyzing the error in terms of the variance-bias decomposition given by

$$E(\hat{\rho}_t(s) - \rho(s))^2 = Var(\hat{\rho}_t(s)) + [bias(\hat{\rho}_t(s))]^2,$$

where $bias(\hat{\rho}_t(s)) = E(\hat{\rho}_t(s) - \rho(s))$.

On the other hand, as is pointed out in [10] (see also [9]), the L_1-criterion is a better error criterion than L_2 for the specific case of density estimation. For instance, L_1 is the natural space for all densities and the integrated absolute error J_t can be recognized visually as the area between the graphs of the density ρ and the estimator $\hat{\rho}_t$. Furthermore, we have the following result, known as Scheffe's Theorem, which is useful to interpret the L_1-criterion.

Theorem D.1 (Scheffe Theorem [9, 10]). *Let ρ and φ be densities. Then*

$$\int |\rho(s) - \varphi(s)| ds = 2 \sup_{D \in \mathscr{B}(\mathfrak{R}^k)} \left| \int_D \rho(s) - \int_D \varphi(s) \right| ds.$$

For two probability measures $\theta, \mu \in \mathbb{P}(\mathfrak{R}^k)$, we define the total variation as

$$\|\theta - \mu\|_{TV} := \sup_{D \in \mathscr{B}(\mathfrak{R}^k)} |\theta(D) - \mu(D)|.$$

If θ and μ have densities ρ and φ respectively, then Scheffe's Theorem yields

$$\|\theta - \mu\|_{TV} = \frac{1}{2} \int |\rho - \varphi| = \frac{1}{2} \|\rho - \varphi\|_{L_1}. \qquad (D.1)$$

Hence, if $\hat{\rho}_t$ is an estimate of the density ρ, from (D.1), with $\hat{\rho}_t$ instead of φ, we have

$$\text{if } \int |\rho(s) - \hat{\rho}_t \rho(s)| \, ds < \varepsilon \text{ then } |\theta(D) - \theta_t(D)| < \frac{\varepsilon}{2} \text{ a.s.,} \qquad \text{(D.2)}$$

for all $D \in \mathscr{B}(\mathfrak{R}^k)$, where $\theta_t(D) = \int_D \hat{\rho}_t(s) ds$.

Another important property of the L_1-criterion is that J_t is invariant under monotone transformations. Indeed, let $T : \mathfrak{R}^k \to \mathfrak{R}^k$ be a one-to-one transformation, and ξ and χ be random variables with densities ρ and φ, respectively. Consider the random variables $\xi^* = T(\xi)$ and $\chi^* = T(\chi)$ with densities ρ^* and φ^*, respectively. Then

$$\|\rho - \varphi\|_{L_1} = \|\rho^* - \varphi^*\|_{L_1}.$$

D.2 Density Estimators

There are several examples of density estimators, for instance the histograms, the kernel estimators, and the projection estimators. Among them, the most understood are the histograms which, in its simplest version, is defined as follows.

For convenience, we assume that $\xi_1, \xi_2, \ldots, \xi_t$ is an i.i.d. sample from a density ρ taking values in a k-rectangle (box) $S \subset \mathfrak{R}^k$. Divide S into disjoint bins or subrectangles $D_1, D_2, \ldots, D_m$ of volume h^k, that is $m \approx \dfrac{1}{h^k}$. The histogram estimate is defined as

$$\hat{\rho}_t(s) = \frac{1}{t} \sum_{i=1}^{t} \frac{1_{[\xi_i \in D_j]}}{h^d}, \quad s \in D_j, j = 1, \ldots, m.$$

Although the histogram is easy to understand and to compute, its applicability is limited in situations where a deeper mathematical analysis is required. For instance, its discontinuity is a limitation when it is necessary to know the derivative of the estimators. However, important density estimators emerge as extension of the histograms, like the kernel estimate on which we will focus.

D.2.1 The Kernel Estimate

A Borel measurable function $K : \mathfrak{R}^k \to \mathfrak{R}$ is a kernel if $K \geq 0$ and $\int_{\mathfrak{R}^k} K(s) ds = 1$.

Definition D.1. Let $\xi_1, \xi_2, \ldots, \xi_t$ be an i.i.d. sample from a density ρ. Given a kernel K, the kernel estimate is defined by

$$\hat{\rho}_t(s) = \frac{1}{t d_t^k} \sum_{i=1}^{t} K \left(\frac{s - \xi_i}{d_t} \right), \quad s \in \mathfrak{R}^k,$$

where d_t is a sequence of positive numbers.

A well-known property of the kernel estimate is its consistency, that is $E(J_t) \to 0$, as $t \to \infty$, which is stated in the following result (see, e.g., [11]).

Theorem D.2. *Let $\hat{\rho}_t$ be a kernel estimate with arbitrary kernel K and $\{d_t\}$ be a sequence such that $d_t \to 0$ and $t d_t^k \to \infty$ as $t \to \infty$. Then $E(J_t) \to 0$ as $t \to \infty$.*

In fact, as is proved in [10], $\hat{\rho}_t$ is strongly consistent, i.e., $J_t \to 0$ in probability, and furthermore all types of convergence to 0 of J_t are equivalent. Specifically we have the following result.

Theorem D.3. *Let $\hat{\rho}_t$ be a kernel estimate with arbitrary kernel K. Then the following statements are equivalent:*
   (a) $J_t \to 0$ in probability as $t \to \infty$.
   (b) $J_t \to 0$ a.s. as $t \to \infty$.
   (c) $J_t \to 0$ exponentially as $t \to \infty$. That is, for all $\varepsilon > 0$ there exist $r, t_0 > 0$ such that

$$P[J_t \geq \varepsilon] \leq e^{-rn}, \quad t \geq t_0.$$

   (d) $d_t \to 0$ and $t d_t^k \to \infty$ as $t \to \infty$.

Remark D.1. Observe that $J_t \leq 2$ a.s. Hence, it is easy to see that if $E(J_t) \to 0$ as $t \to \infty$, then, for any $q > 0$, $E(J_t)^q \to 0$.

Analyzing the consistency of the kernel estimate $\hat{\rho}_t$ with the mean integrated absolute error $E(J_t)$, we can obtain its rate of convergence. If this is the case, it is necessary to impose conditions on the density ρ. For simplicity, we present the result when $k = 1$, which is proved, for instance, in [11] (see also [10]).

Theorem D.4. *Let $\hat{\rho}_t$ be a kernel estimate where K is a kernel such that $\int s^2 K(s)ds < \infty$ and $K(s) = K(-s)$, $s \in \mathfrak{R}$. In addition, $\{d_t\}$ is a sequence such that $d_t \to 0$ and $t d_t^k \to \infty$ as $t \to \infty$. Let ρ be a density satisfying the following conditions:*
   (i) $\int |\rho''| < \infty$;
   (ii) $\int \sqrt{\rho^*} < \infty$, where $\rho^*(s) = \sup_{s-1 \leq y \leq s+1} \rho(y)$.
Then $E(J_t) = O(t^{-2/5})$ as $t \to \infty$.

Remark D.2. In the same sense of the Remark D.1, if $E(J_t) = O(t^{-\gamma})$ as $t \to \infty$ for some $\gamma > 0$, then, for any $q > 1$,

$$E(J_t)^q = O(t^{-\gamma}) \text{ as } t \to \infty. \tag{D.3}$$

Indeed, if $q > 1$,

$$E(J_t)^q = E\left[(J_t)(J_t)^{q-1}\right] \leq 2^{q-1} E(J_t),$$

which, because $E(J_t) = O(t^{-\gamma})$ as $t \to \infty$, implies (D.3).

Depending on their features, it is expected that there are densities that are more difficult to estimate than others, so the efficiency of the same kernel density estimator may be different. In addition, the accuracy and good behavior of the kernel density estimator depend on other factors such as the choice of the kernel K and the bandwidth h. A thorough analysis of these issues, as well as examples of estimators under several context, can be found, for instance, in [9–11, 27, 74, 75].

D.2.2 Projection Estimates

In certain situations, beside the usual statistical properties, the estimators need to satisfy additional properties. For example, it is commonly required that the estimate $\hat{\rho}_t$ and the density ρ have the same functional properties. This case can be addressed in the following manner.

Let $\mathscr{D} \subset L_1(\mathfrak{R}^k)$ be a class of densities containing ρ. Let $\hat{\rho}_t$ be an arbitrary density estimate. We define the *projection estimate* as the projection ρ_t of $\hat{\rho}_t$ onto the class $\mathscr{D}$, that is

$$\|\rho_t - \hat{\rho}_t\|_{L_1} = \inf_{\varphi \in \mathscr{D}} \|\varphi - \hat{\rho}_t\|_{L_1}.$$

The following result provides conditions ensuring the existence of the projection estimate ρ_t.

Theorem D.5. *Let $\mathscr{D} \subset L_1(\mathfrak{R}^k)$ be a closed and convex class of densities and $\hat{\rho}_t$ be an arbitrary estimate. Then there exists the projection estimate $\rho_t \in \mathscr{D}$, which is defined as*

$$\rho_t = \underset{\varphi \in \mathscr{D}}{\operatorname{argmin}} \|\varphi - \hat{\rho}_t\|_{L_1}. \tag{D.4}$$

The proof of this result for L_q-spaces can be found in [45]. For the case of densities see [11].

Remark D.3. Observe that from (D.4) we have

$$\begin{aligned}
\|\rho_t - \rho\|_{L_1} &\leq \|\rho_t - \hat{\rho}_t\|_{L_1} + \|\hat{\rho}_t - \rho\|_{L_1} \\
&\leq \|\hat{\rho}_t - \rho\|_{L_1} + \|\hat{\rho}_t - \rho\|_{L_1} \\
&= 2\|\hat{\rho}_t - \rho\|_{L_1}.
\end{aligned}$$

Hence, the consistency of the projection estimate ρ_t is inherited from the consistency of $\hat{\rho}_t$, as well as the rate of convergence.

D.2.3 The Parametric Case

Consider the case in which the unknown density ρ belongs to a parametric family of densities $\{\rho_\lambda : \lambda \in \Lambda\}$, where Λ is the parameter set. So, when estimating the unknown parameter we obtain a density estimate whose properties will depend on the parametric estimation method used. For instance, under some regularity conditions, the maximum likelihood method yields consistent estimates (see, e.g., [68]). Below we will illustrate a particular case.

Let $\varphi : \Re \to \Re^+$ be a measurable function such that $\int \varphi < \infty$. We assume that the unknown density ρ_λ is of the form

$$\rho_\lambda(s) = \frac{\varphi(s)}{\int_\lambda^\infty \varphi(s)ds} 1_{[\lambda,\infty)}(s), \quad \lambda \in \Lambda,$$

where Λ is a subset of $\Re^+$. It can be proven that $\lambda_t := \min\{\xi_1, \xi_2, \ldots, \xi_t\}$ is the maximum likelihood estimate of λ and

$$\lambda_t \to \lambda \text{ a.s., as } t \to \infty. \tag{D.5}$$

Let F_λ be the distribution function corresponding to the density ρ_λ, and define

$$\rho_t(s) = \rho_{\lambda_t}(s) = \frac{\varphi(s)}{\int_{\lambda_t}^\infty \varphi(s)ds} I_{[\lambda_t,\infty)}(s).$$

Observe that

$$|\rho_\lambda(s) - \rho_t(s)| = \begin{cases} \rho_\lambda(s), & \text{if } \lambda < s \leq \lambda_t; \\ \rho_t(s) - \rho_\lambda(s), & \text{if } \quad s > \lambda_t. \end{cases}$$

Hence

$$J_t = \int_\lambda^\infty |\rho_\lambda(s) - \rho_t(s)| ds = 2\int_\lambda^{\lambda_t} \rho_\lambda(s)ds = 2F_\lambda(\lambda_t).$$

Thus, from (D.5), $J_t \to 0$ a.s., as $t \to \infty$, that is ρ_t is (strong) consistent. Furthermore

$$E(J_t) = 2E\left[F_\lambda(\lambda_t)\right] = 2\int_\lambda^\infty F_\lambda(s)\rho_{\lambda_t}(s)ds$$

$$= 2\int_\lambda^\infty tF_\lambda(s)(1 - F_\lambda(s))^{t-1}\rho_\lambda(s)ds \tag{D.6}$$

$$= \frac{2}{t+1}.$$

In (D.6) we have used the fact that the distribution function of λ_t is $F_{\lambda_t}(s) = 1 - (1 - F_\lambda(s))^t$, which yields $\rho_{\lambda_t}(s) = t(1 - F_\lambda(s))^{t-1}\rho_\lambda(s)$. Therefore,

$$E(J_t) = O(t^{-\gamma}) \text{ for } \gamma \in (0, 1].$$

References

1. Aliprantis, C.D., Border, K.C.: Infinite Dimensional Analysis. Springer, Berlin (1999)
2. Altman, E., Shwartz, A.: Adaptive control of constrained Markov chains: criteria and policies. Ann. Oper. Res. **28**, 101–134 (1991)
3. Ash, R.: Real Analysis and Probability. Academic, New York (1972)
4. Berge, E.: Topological Spaces. Macmillan, New York (1963)
5. Bertsekas, D.P., Shreve, S.E.: Stochastic Optimal Control: The Discrete Time Case. Academic, New York (1978)
6. Billingsley, P., Topsoe, F.: Uniformity in weak convergence. Z. Wahrsch. Verw. Geb. **7**, 1–16 (1967)
7. Cavazos-Cadena, R.: Nonparametric adaptive control of discounted stochastic systems with compact state space. J. Optim. Theory Appl. **65**, 191–207 (1990)
8. Chang, H.S.: Perfect information two-person zero-sum Markov games with imprecise transition probabilities. Math. Methods Oper. Res. **64**, 235–351 (2006)
9. Devroye, L.: A Course in Density Estimation. Birkhäuser, Boston (1987)
10. Devroye, L., Györfi, L.: Nonparametric Density Estimation. The L_1 View. Wiley, New York (1985)
11. Devroye, L., Lugosi, G.: Combinatorial Methods in Density Estimation. Springer, New York (2001)
12. Dudley, R.M.: The speed of mean Glivenko-Cantelli convergence. Ann. Math. Stat. **40**, 40–50(1969)
13. Dynkin, E.B., Yushkevich, A.A.: Controlled Markov Processes. Springer, New York (1979)
14. Engwerda, J.: LQ Dynamic Optimization and Differential Games. Wiley, Hoboken (2005)
15. Fan, K.: Minimax theorems. Proc. Nat. Acad. Sci. U. S. A. **39**, 42–47 (1953)
16. Fernández-Gaucherand, E.: A note on the Ross-Taylor Theorem. Appl. Math. Comput. **64**, 207–212 (1994)
17. Filar, J., Vrieze, K.: Competitive Markov Decision Processes. Springer, New York (1997)
18. Gaenssler, P., Stute, W.: Empirical processes: a survey for i.i.d. random variables. Ann. Probab. **7**, 193–243 (1979)
19. Ghosh, M.K., Bagachi, A.: Stochastic games with average payoff criterion. Appl. Math. Optim. **38**, 283–301 (1998)
20. Ghosh, M.K., McDonald, D., Sinha, S.: Zero-sum stochastic games with partial information. J. Optim. Theory Appl. **121**, 99–118 (2004)
21. González-Trejo, T.J., Hernández-Lerma, O., Hoyos-Reyes, L.F.: Minimax control of discrete-time stochastic systems. SIAM J. Control Optim. **41**, 1626–1659 (2003)
22. Gordienko, E.I.: Adaptive strategies for certain classes of controlled Markov processes. Theory Probab. Appl. **29**, 504–518 (1985)

23. Gordienko, E.I., Hernández-Lerma, O.: Average cost Markov control processes with weighted norms: existence of canonical policies. Appl. Math. **23**, 119–218 (1995)
24. Gordienko, E.I., Hernández-Lerma, O.: Average cost Markov control processes with weighted norms: value iteration. Appl. Math. **23**, 219–237 (1995)
25. Gordienko, E.I., Minjárez-Sosa, J.A.: Adaptive control for discrete-time Markov processes with unbounded costs: discounted criterion. Kybernetika **34**, 217–234 (1998)
26. Gordienko, E.I., Minjárez-Sosa, J.A.: Adaptive control for discrete-time Markov processes with unbounded costs: average criterion. ZOR- Math. Methods Oper. Res. **48**, 37–55 (1998)
27. Hasminskii, R., Ibragimov, I.: On density estimation in the view of Kolmogorov's ideas in approximation theory. Ann. Stat. **18**, 999–1010 (1990)
28. Hernández-Lerma, O.: Adaptive Markov Control Processes. Springer, New York (1989)
29. Hernández-Lerma, O., Cavazos-Cadena, R.: Density estimation and adaptive control of Markov processes: average and discounted criteria. Acta Appl. Math. **20**, 285–307 (1990)
30. Hernández-Lerma, O., Lasserre, J.B.: Discrete-Time Markov Control Processes: Basic Optimality Criteria. Springer, New York (1996)
31. Hernández-Lerma, O., Lasserre, J.B.: Further Topics on Discrete-Time Markov Control Processes. Springer, New York (1999)
32. Hernández-Lerma, O., Lasserre, J.B.: Zero-sum stochastic games in Borel spaces: average payoff criteria. SIAM J. Control Optim. **39**, 1520–1539 (2001)
33. Hilgert, N., Minjárez-Sosa, J.A.: Adaptive policies for time-varying stochastic systems under discounted criterion. Math. Methods Oper. Res. **54**, 491–505 (2001)
34. Hilgert, N., Minjárez-Sosa, J.A.: Limiting average cost adaptive control problem for time – varying stochastic systems. Bol. Soc. Mat. Mexicana **9**, 197–212 (2003)
35. Hilgert, N., Minjárez-Sosa, J.A.: Adaptive control of stochastic systems with unknown disturbance distribution: discounted criteria. Math. Methods Oper. Res. **63**, 443–460 (2006)
36. Himmelberg, C.J., Van Vleck, F.S.: Multifunctions with values in a space of probability measures. J. Math. Anal. Appl. **50**, 108–112 (1975)
37. Himmelberg, C.J., Parthasarathy, T., Van Vleck, F.S.: Optimal plans for dynamic programming problems. Math. Oper. Res. **1**, 390–294 (1976)
38. Hinderer, K.: Foundations of Non-stationary Dynamic Programming with Discrete Time Parameter. Lecture Notes in Operations Research and Mathematical Systems, vol. 33. Springer, Berlin (1970)
39. Jaśkiewicz, A.: A fixed point approach to solve the average cost optimality equation for semi-Markov decision processes with Feller transition probabilities. Commun. Stat. Theory Methods **36**, 2559–2575 (2007)
40. Jaśkiewicz, A.: Zero-sum ergodic semi-Markov games with weakly continuous transition probabilities. J. Optim. Theory Appl. **141**, 321–347 (2009)

41. Jaśkiewicz, A.: On a continuous solution to the Bellman-Poisson equation in stochastic games. J. Optim. Theory Appl. **145**, 451–458 (2010)
42. Jaśkiewicz, A., Nowak, A.: Zero-sum ergodic stochastic games with Feller transition probabilities. SIAM J. Control Optim. **45**, 773–789 (2006)
43. Jaśkiewicz, A., Nowak, A.: Approximation of noncooperative semi-Markov games. J. Optim. Theory Appl. **131**, 115–134 (2006)
44. Jaśkiewicz, A., Nowak, A.: On the optimality equation for average cost Markov control processes with Feller transitions probabilities. J. Math. Anal. Appl. **316**, 495–509 (2006)
45. Köthe, G.: Topological Vector Spaces I. Springer, New York (1969)
46. Krausz, A., Rieder, U.: Markov games with incomplete information. Math. Methods Oper. Res. **46**, 263–279 (1997)
47. Küenle, H.U.: On Markov games with average reward criterion and weakly continuous transition probabilities. SIAM J. Control Optim. **46**, 2156–2168 (2007)
48. Luque-Vásquez, F.: Zero-sum semi-Markov games in Borel spaces: discounted and average payoff. Bol. Soc. Mat. Mexicana **8**, 227–241 (2002)
49. Luque-Vázquez, F., Minjárez-Sosa, J.A.: A note on the σ-compactness of sets of probability measures on metric spaces. Statist. Probab. Lett. **84**, 212–214 (2014)
50. Luque-Vázquez, F., Minjárez-Sosa, J.A.: Empirical approximation in Markov games under unbounded payoff: discounted and average criteria. Kybernetika **53**, 694–716 (2017)
51. Meyn, S.P., Tweedie, R.L.: Markov Chains and Stochastic Stability. Springer, London (1993)
52. Minjárez-Sosa, J.A.: Nonparametric adaptive control for discrete-time Markov processes with unbounded costs under average criterion. Appl. Math. **26**, 267–280 (1999)
53. Minjárez-Sosa, J.A.: Average optimality for adaptive Markov control processes with unbounded costs and unknown disturbance distribution. In: Zhenting, H., Filar, J.A., Chen, A. (eds.) Markov Processes and Controlled Markov Chains, Chap. 7. Kluwer, Dordrecht (2002)
54. Minjárez-Sosa, J.A.: Approximation and estimation in Markov control processes under a discounted criterion. Kybernetika **40**, 681–690 (2004)
55. Minjárez-Sosa, J.A.: Empirical estimation in average Markov control processes. Appl. Math. Lett. **21**, 459–464 (2008)
56. Minjárez-Sosa, J.A., Luque-Vásquez, F.: Two person zero-sum semi-Markov games with unknown holding times distribution on one side: discounted payoff criterion. Appl. Math. Optim. **57**, 289–305 (2008)
57. Minjárez-Sosa, J.A., Vega-Amaya, O.: Asymptotically optimal strategies for adaptive zero-sum discounted Markov games. SIAM J. Control Optim. **48**, 1405–1421 (2009)
58. Minjárez-Sosa, J.A., Vega-Amaya, O.: Optimal strategies for adaptive zero-sum average Markov games. J. Math. Anal. Appl. **402**, 44–56 (2013)

59. Najim, K., Poznyak, A.S., Gómez, E.: Adaptive policy for two finite Markov chains zero-sum stochastic game with unknown transition matrices and average payoffs. Automatica **37**, 1007–1018 (2001)
60. Neyman, A., Sorin, S.: Stochastic Games and Applications. Kluwer, Dordrecht (2003)
61. Nowak, A.: Measurable selection theorems for minimax stochastic optimization problems. SIAM J. Control Optim. **23**, 466–476 (1985)
62. Parthasarathy, K.R.: Probability Measures on Metric Spaces. Academic, New York (1967)
63. Prieto-Rumeau, T., Lorenzo, J.M.: Approximation of zero-sum continuous-time Markov games under the discounted payoff criterion. TOP **23**, 799–836 (2015)
64. Ranga Rao, R.: Relations between weak and uniform convergence of measures with applications. Ann. Math. Stat. **33**, 659–680 (1962)
65. Shapley, L.S.: Stochastic games. Proc. Nat. Acad. Sci. U. S. A. **39**, 1095–1100 (1953)
66. Schäl, M.: Conditions for optimality and for the limit of n-stage optimal policies to be optimal. Z. Wahrs. Verw. Gerb. **32**, 179–196 (1975)
67. Schäl, M.: Estimation and control in discounted stochastic dynamic programming. Stochastics **20**, 51–71 (1987)
68. Serfling, R.J.: Approximation Theorems of Mathematical Statistics. Wiley, Hoboken (2002)
69. Shimkin, N., Shwartz, A.: Asymptotically efficient adaptive strategies in repeated games. Part I: certainty equivalence strategies. Math. Oper. Res. **20**, 743–767 (1995)
70. Shimkin, N., Shwartz, A.: Asymptotically efficient adaptive strategies in repeated games. Part II: asymptotic optimality. Math. Oper. Res. **21**, 487–512 (1996)
71. Van der Vaart, A.E., Wellner, J.A.: Weak Convergence and Empirical Processes. Springer, New York (1996)
72. Van Nunen, J.A.E.E., Wessels, J.: A note on dynamic programming with unbounded rewards. Manag. Sci. **24**, 576–580 (1978)
73. Vega-Amaya, O.: Zero-sum average semi-Markov games: fixed point solutions of the Shapley equation. SIAM J. Control Optim. **42**, 1876–1894 (2003)
74. Wand, M.P., Devroye, L.: How easy is a given density to estimate? Comput. Stat. Data Anal. **16**, 311–323 (1993)
75. Wand, M.P., Jones, M.C.: Kernel Smoothing. Chapman and Hall, London (1995)
76. Wessels, J.: Markov programming by successive approximations with respect to weighted supremum norms. J. Math. Anal. Appl. **58**, 326–335 (1977)
77. Yakowitz, S.: Dynamic programming applications in water resources. Water Resour. Res. **18**, 673–696 (1982)
78. Yeh, W.W-G.: Reservoir management and operations models: a state-of-the-art review. Water Resour. Res. **21**, 1797–1818 (1985)

Index

Printed in the United States
By Bookmasters